Microbiology
of
Solid Waste

The Microbiology of
EXTREME AND UNUSUAL ENVIRONMENTS

————————— *SERIES EDITOR* —————————
RUSSELL H. VREELAND

Titles in the Series

The Biology of Halophilic Bacteria
Russell H. Vreeland and Lawrence Hochstein

The Microbiology of Deep-Sea Hydrothermal Vents
David M. Karl

Microbiology of Solid Waste
Anna C. Palmisano and Morton A. Barlaz

The Microbiology of the Terrestrial Subsurface
Penny S. Amy and Dana L. Haldeman

Microbiology
of
Solid Waste

Edited by

Anna C. Palmisano
Morton A. Barlaz

CRC Press
Taylor & Francis Group
Boca Raton London New York

CRC Press is an imprint of the
Taylor & Francis Group, an **informa** business

CRC Press
Taylor & Francis Group
6000 Broken Sound Parkway NW, Suite 300
Boca Raton, FL 33487-2742

© 1996 by Taylor & Francis Group, LLC
CRC Press is an imprint of Taylor & Francis Group, an Informa business

No claim to original U.S. Government works

ISBN-13: 978-0-8493-8361-8 (hbk)

Visit the Taylor & Francis Web site at
http://www.taylorandfrancis.com
and the CRC Press Web site at
http://www.crcpress.com

Cover Designer: Denise Craig
Typesetting: Roy Barnhill

Library of Congress Card Number 96-18240

Library of Congress Cataloging-in-Publication Data

Microbiology of solid waste / edited by Anna C. Palmisano, Morton A.
    Barlaz.
        p.   cm. -- (Microbiology of extreme and unusual environments)
    Includes bibliographical references and index.
    ISBN 0-8493-8361-7
    1. Sanitary microbiology. 2. Hazardous wastes--Microbiology.
3. Refuse and refuse disposal--Biodegradation. 4. Integrated solid
waste management.   I. Palmisano, Anna C.   II. Barlaz, Morton A.
III. Series.
QR48.M53   1996
576'. 16--dc20   96-18240
                                                                      CIP

Preface

The *Microbiology of Solid Waste* provides an overview of this multidisciplinary field for microbiologists, environmental engineers, and solid waste managers. Solid waste microbiology had its origins in public health microbiology in the early 1960s, when researchers emphasized the potential for solid waste to harbor pathogenic microorganisms. Interest in solid waste was renewed in the 1980s, as society became more interested in the broader environmental impacts of solid waste and how biological processes might be utilized to minimize such impacts. Research emerged on solid waste decomposition under anaerobic and aerobic conditions, in addition to the development of biodegradable plastics and new tools for the detection of pathogens using molecular approaches. This book attempts to summarize this recent growth in our understanding of the microbial ecosystems involved in solid waste decomposition.

To our knowledge, this is the first attempt to bring together information on microbial communities associated with decomposing solid waste in landfills, compost facilities, and anaerobic digesters, including potential pathogens, and approaches for testing the biodegradability of new materials that will ultimately be managed as solid waste. Each author has identified gaps in our current knowledge and opportunities for future research.

The need to consider solid waste management as an integrated system of collection, recycling, treatment, and disposal is discussed in the introduction. It is our hope that this new understanding of solid waste microbiology will contribute to integrated approaches for safe and economical solid waste management. Dr. Palmisano thanks the Office of Naval Research for release time for the editing of this book. Dr. Barlaz thanks the Environmental Protection Agency, the National Science Foundation, the Procter & Gamble Company, the American Society for Testing Materials, Waste Management of North America, and S.C. Johnson Wax & Son for the support of his research on solid waste decomposition. These organizations have helped lead the discipline of solid waste microbiology forward.

Anna C. Palmisano
Morton A. Barlaz

The Editors

Anna Palmisano, Ph.D., is the Acting Director of the Biological and Biomedical Science and Technology Division of the Office of Naval Research. She received her B.S. in Microbiology from the University of Maryland (*summa cum laude*) and her M.S. and Ph.D. in Biology from the University of Southern California. Dr. Palmisano was a National Research Council postdoctoral fellow at NASA-Ames Laboratory in Planetary Biology, and she worked as an environmental microbiologist for the Procter & Gamble Company.

Dr. Palmisano is a member of the American Society for Microbiology, American Society for Limnology and Oceanography, Phycological Society of America, and the Association of Women in Science. She serves on the editorial board for the *Journal of Microbiological Methods*. Her research interests have included the microbial ecology of solid waste decomposition, biodegradation of xenobiotics in fresh water, microbial mat communities, and psychrophilic microorganisms.

Morton A. Barlaz, Ph.D., is an Associate Professor of Civil Engineering at North Carolina State University. He received his B.S. in Chemical Engineering from the University of Michigan and M.S. and Ph.D. degrees in Civil and Environmental Engineering from the University of Wisconsin. Dr. Barlaz worked as an environmental engineer in both industry and government before beginning his academic career.

Dr. Barlaz is a member of the American Society for Microbiology, the International Solid Waste Management Association, the American Society of Civil Engineers, Sigma Xi, and the Water Environment Federation. Dr. Barlaz serves on the editorial board of *Waste Management and Research* and the solid waste committee of the American Society of Civil Engineers. Dr. Barlaz is conducting research on the anaerobic decomposition of wastes in landfills, alternate strategies for the management of municipal solid waste in consideration of both economics and environmental burdens, and bioremediation of contaminated aquifers.

Contributors

Anna C. Palmisano, Ph.D.
Program Officer
Biological Science and Technology
  Program
Office of Naval Research
Department of the Navy
Arlington, Virginia

Morton A. Barlaz, Ph.D., P.E.
Associate Professor
Department of Civil Engineering
North Carolina State University
Raleigh, North Carolina

David P. Chynoweth, Ph.D.
Professor
Department of Agricultural and
  Biological Engineering
University of Florida
Gainesville, Florida

Charles P. Gerba, Ph.D.
Professor
Department of Soil, Water Science,
  and Environmental Science
University of Arizona
Tucson, Arizona

Bradley N. Johnson, Ph.D.
Paper Product Development
The Procter & Gamble Company
Cincinnati, Ohio

Frederick C. Miller, Ph.D.
Cabot, Pennsylvania

Charles A. Pettigrew, Ph.D.
Paper Product Development
The Procter & Gamble Company
Cincinnati, Ohio

Pratap Pullammanappallil, Ph.D.
CRC for Waste Management and
  Pollution Control Ltd.
Department of Chemical Engineering
The University of Queensland
St. Lucia, Queensland
Australia

Contents

Chapter 1

Introduction to Solid Waste Decomposition 1
Palmisano and Barlaz

Chapter 2

Microbiology of Solid Waste Landfills 31
Barlaz

Chapter 3

Anaerobic Digestion of Municipal Solid Wastes 71
Chynoweth and Pullammanappallil

Chapter 4

Composting of Municipal Solid Waste
and Its Components 115
Miller

Chapter 5

Microbial Pathogens in Municipal Solid Waste 155
Gerba

Chapter 6

Testing the Biodegradability of Synthetic
Polymeric Materials in Solid Waste 175
Pettigrew and Johnson

Index 215

1

Introduction to Solid Waste Decomposition

Anna C. Palmisano and Morton A. Barlaz

CONTENTS

1.1 Introduction 2
    1.1.1 Objective of Book 2
    1.1.2 Refuse as a Niche for Microbial Growth 2
    1.1.3 Microbiology and Economics 3

1.2 Municipal Solid Waste Composition 3
    1.2.1 Sortable Waste Components 4
    1.2.2 Chemical Contaminants in MSW 5

1.3 Integrated Solid Waste Management 6

1.4 Biological Decomposition 9
    1.4.1 Landfill Microbial Communities 9
    1.4.2 Anaerobic Digestion of MSW 14
    1.4.3 Composting of MSW 17

1.5 Socioeconomic Issues 18
    1.5.1 Public Health Concerns 19
    1.5.2 Contribution to Greenhouse Gases 20
    1.5.3 Biodegradability of Polymeric Materials in MSW 21

1.6 Decision-Making Tools for Solid Waste Management 22

1.7 Research Gaps and Opportunities 23
    1.7.1 Community Analysis by Molecular and Biochemical Methods 23
    1.7.2 The Role of Bactivory and Lysis in Refuse-Associated
            Microbial Communities 24
    1.7.3 Surface Analysis of Microbial Communities 24
    1.7.4 Designing Polymers for Hydrolysis 25

1.8 Conclusions 25

References 25

0-8493-8361-7/96/$0.00+$.50
© 1996 by CRC Press, Inc.

1

1.1 INTRODUCTION

1.1.1 Objective of Book

The objective of this book is to provide a basic understanding of the microbial communities associated with the decomposition of municipal solid waste (MSW). Historically, decomposition by microorganisms has received little attention relative to the large amounts of MSW generated. This has begun to change as society struggles to deal with the ever growing volumes of solid waste. In this book, we will limit our discussions to MSW (see Section 1.2). Construction debris, wastewater treatment sludges, and other materials some-times managed with MSW present unique problems for decomposition and are beyond the purview of this discussion. In later chapters, the decomposition of MSW under anaerobic conditions in landfills and digesters and under aerobic conditions in composting will be discussed in some detail. Chapters are devoted also to pathogens that may be associated with MSW and methods to ascertain biodegradability of materials under typical conditions of MSW disposal. In this introductory chapter, we will provide a brief overview of these topics and place them in the larger context of societal impact.

1.1.2 Refuse as a Niche for Microbial Growth

Solid waste provides both substrate and substratum for the growth and succession of diverse microbial communities. Landfills, composting facilities, and anaerobic digesters are unique ecosystems for the development of complex microbial communities. Many aspects of refuse are conducive to the growth of microorganisms. Surfaces are available for colonization, organic and inorganic nutrients are abundant, moisture is usually adequate (at least in portions of the waste), and temperatures are often elevated with respect to surface soil environments. Since typical refuse substituents such as paper, food, and yard waste are polymeric in nature, substrates for microbial growth include cellulose, starch, protein, and hemicellulose (Jones and Grainger, 1983; Barlaz et al., 1990).

The heterogeneity created by the physical environment of refuse poses challenges to microbes and microbiologists alike. Watson-Craik and Jones (1995) wrote that, "Each sample of refuse itself ... constitutes an exceptionally heterogeneous environment which comprises a wide range of organic molecules of both natural and xenobiotic origin, some or all of which serve as substrates for microbial growth, which are irregularly distributed in medium composed of surfaces of a varying nature and sporadically bathed in a fluid of uncertain composition." For the microbes, refuse heterogeneity can promote diversity by creating microniches that allow otherwise incompatible microbial processes to occur in close proximity. For the researcher, however, the heterogeneity of refuse creates difficulties in obtaining representative samples. Some microbiologists have chosen to shred refuse material prior to incubation or assaying for activity (Bookter and Ham, 1982). Others have sieved out large, intractible refuse objects such as shoes and tires prior to sample analysis (Suflita et al., 1992). In either case, the heterogeneity of MSW on both large and small scales necessitates the collection of a large number of samples per site. Moreover, the solid waste stream may vary considerably with climate, season, disposal practices, and a wide range of socioeconomic factors (Tchobanoglous et al., 1993). For example, in some areas in the U.S., yard waste is separated from MSW for composting, thus changing the milieu for microbial growth in the residual MSW and creating a unique ecosystem for the compost. Moisture levels in landfills in the arid southwestern U.S. are much lower than those in the southeast, for example, and this can have a profound effect on rates of refuse biodegradation.

Upon disposal, solid waste supports a succession of microbial communities. The diversity of microorganisms that colonize refuse and grow under the conditions of solid waste disposal in landfills, digesters, and composters is only beginning to be appreciated. Microbial communities in landfills and anaerobic digesters include hydrolytic and fermentative bacteria, acetogens, methanogens, sulfate reducers, and protozoa. Compost fosters the growth of fungi and bacteria (including actinomycetes and thermophiles). The limitations imposed by the need to isolate bacteria on agar plates still constrain our knowledge of the wealth of microbial diversity. New methods of molecular analysis are confirming that less than 1% of the microbes present in natural environments are culturable (Amann et al., 1995). These molecular methods are just beginning to be applied to refuse samples and will undoubtedly greatly expand our knowledge of the microbial diversity associated with decomposing MSW.

1.1.3 Microbiology and Economics

Optimization of microbial activity in decomposing refuse has important economic implications. Tapping of methane from landfills and anaerobic digesters can provide a marketable endproduct in the form of energy. Moreover, formation of benign products means minimization of nuisance compounds such as volatile fatty acids or potentially toxic substrates. In composting and anaerobic digestion, microbial decomposition results in a reduction in the volume of refuse; moreover, the residues may have economic value as a soil amendment. Attenuation of pathogens by high temperatures achieved in microbial decomposition is another benefit derived from microbial activities.

1.2 MUNICIPAL SOLID WASTE COMPOSITION

Solid waste is characterized by the presence of an abundance of degradable carbon, and solid waste microbiology is influenced by the nature of this carbon. Thus, it is important to understand the composition of the waste which may undergo biodegradation. In this section, we define MSW and present data on its composition.

1.2.1 Sortable Waste Components

The U.S. Environmental Protection Agency (USEPA) has published a characterization of MSW as well as trends in MSW management for nearly 20 years. In a discussion of MSW, it is critical to understand the wastes that are included and excluded in its definition. As defined by the USEPA (1994a), MSW includes "wastes such as durable goods, nondurable goods, containers and packaging, food scraps, yard trimmings, and miscellaneous inorganic wastes from residential, commercial, institutional, and industrial sources." Specifically excluded from MSW are construction and demolition wastes, water and wastewater treatment sludges, combustion ash, and industrial process waste. Thus, MSW represents a subset of the solid waste that must be managed by a particular community. While this book largely focuses on MSW, some of these other wastes may be handled in a similar manner.

MSW composition is presented by both weight and volume percent in Table 1. While weight is more easily measured and is the more traditional measure, volume data are important because landfills reach capacity by volume. The weight and volume data in Table 1 are not directly comparable, because weight data are based on an estimate of MSW as generated, while volume data are based on MSW buried in landfills after removal of that material which is recycled, combusted, or composted. Moreover, given the difficulty in estimating the space occupied by individual components after burial, there is greater uncertainty associated with the volume data. Paper and paperboard dominate MSW on both weight and volume bases. The most significant difference between the weight and volume data is for plastics, which represent 9.3 and 23.9% of the weight generated and the volume landfilled, respectively. Yard trimmings, which are increasingly being composted rather than landfilled, represented about 15.9% of MSW generated and 8.1% of the volume of MSW landfilled in 1993. The inherently biodegradable fraction of MSW (paper and paperboard, food waste, and yard trimmings) comprises approximately 60% of the weight of MSW generated and 41% of the volume landfilled.

TABLE 1   Composition of MSW in 1993

Component	Weight as Generated (%)	Volume as Landfilled (%)
Paper and paperboard	37.6	30.2
Glass	6.6	2.2
Metal	8.3	10.3
Plastic	9.3	23.9
Wood	6.6	6.8
Food	6.7	3.2
Yard trimmings	15.9	8.1
Other (rubber, textiles, etc.)	9.0	15.3

Source: USEPA. 1994a. Characterization of Municipal Solid Waste in the United States: 1994 Update. EPA/530-R-94-042. U.S. Environmental Protection Agency, Washington, D.C.

1.2.2 Chemical Contaminants in MSW

The MSW composition data presented in Table 1 includes a category called "other" which includes a number of small quantity wastes such as diapers, pet waste, and hazardous chemicals. Sources of hazardous waste in landfills include household hazardous waste, waste disposed of by small quantity generators of hazardous waste who are exempted from hazardous waste disposal regulations, and hazardous waste illegally combined with MSW (Reinhart, 1993). Examples of household hazardous wastes and the chemicals which they contain are presented in Table 2. In addition to hazardous wastes as a source of toxic chemicals, Wilkins (1994) identified 90 volatile organic compounds in the headspace in waste collection vehicles, many of which were thought to be byproducts of microbiological activity or components of nonhazardous packaging materials. Once volatile organic compounds enter the MSW stream, they may be released to the environment by a number of mechanisms. The most common mechanism is volatilization, which may occur during composting, anaerobic digestion, or burial of MSW. The presence of toxic trace organics in landfill gas is well documented, and the list includes petroleum hydrocarbons (benzene, toluene, ethylbenzene, and xylenes, or BTEX), chlorinated aliphatic hydrocarbons (CAHs), ketones, and others (USEPA, 1991; Tchobanoglous et al., 1993). In the case of CAHs, trichloroethylene and perchloroethylene may be converted to less chlorinated compounds, including dichloroethylene, vinyl chloride (VC), and ethylene by an anaerobic microbial process known as *reductive dehalogenation* (Mohn and Tiedje, 1992). In this process, the CAH serves as an electron acceptor, and a hydrogen is substituted for a chlorine moiety. Vinyl chloride, which is highly volatile, is a common constituent of landfill gas. In a survey of landfills in California, the median concentration of vinyl chloride was 1150 ppb (Tchobanoglous et al., 1993).

TABLE 2   Household Products Containing Toxic Chemicals

Product Function	Hazardous Chemical
Automotive gasket remover, paint remover, glues	Methylene chloride
Tire sealant	Perchloroethylene
Glues	Trichloroethylene
Paint primer, paint, automotive fuel	Toluene
Paint remover, paint	Xylene
Automotive fuel	Benzene, ethylbenzene
Batteries	Lead, cadmium

Trace organic compounds, including those listed above, are also commonly detected in landfill leachate (Christensen et al., 1994). Heavy metals also may be present in landfill leachate, although concentrations typically are not of concern (Rhew and Barlaz, 1995). Of course, where specific heavy metal-bearing wastes are combined with MSW, their concentrations in leachate may be elevated.

The presence of heavy metals and trace organics is of concern in treatment processes such as anaerobic digestion and composting where there may be an attempt to apply the residual and nondegraded solid waste to the terrestrial environment as a soil amendment. For this reason, there is typically substantial preprocessing of the MSW prior to composting or anaerobic digestion to remove metals and other nondegradable constituents. In some cases, only a specific fraction of MSW (such as yard trimmings and food waste) is collected for these treatments to minimize the potential for significant concentrations of trace organics and heavy metals in the soil amendment product.

1.3 INTEGRATED SOLID WASTE MANAGEMENT

In 1993, MSW generation was estimated to be 2 kg per person per day (USEPA, 1994a). Approximately 62.4% of the 206.9 million tons (188.1 million metric tons) of MSW generated were landfilled, 15.9% were combusted, and 21.7% were recycled or composted. There have been substantial increases in the fraction of the waste stream composted and recycled over the past five years, while the fraction combusted has remained nearly constant. Despite increases in recycling and composting, MSW generation rates have increased. Thus, the total mass of MSW buried in landfills is not decreasing significantly. The role of various treatment and disposal options in solid waste management is described below.

The last decade has witnessed a dramatic increase in the amount of attention given to the manner in which solid waste is managed (O'Leary et al., 1988). Historically, MSW was collected in a single collection vehicle and transported either to a landfill for burial or to a combustion facility for volume reduction followed by burial of the resultant ash. Where long distances were involved, transfer stations may have been employed. Over the last decade, policy directives requiring more recycling, banning yard trimmings from landfills, and requiring landfill designs that restrict leachate release have resulted in increased costs for solid waste management. Subsequently, solid waste managers have recognized the importance of developing integrated solid waste management strategies.

Municipal solid waste management may include a number of processes including collection, separation, treatment, and disposal. Figure 1 presents a number of alternatives for MSW collection, including mixed refuse collection, separate collection of recyclables to facilitate recycling, and separate collection of yard trimmings to facilitate composting. MSW components that are commonly recycled include aluminum and ferrous metal cans, newsprint, milk and water containers (translucent high density polyethylene), carbonated beverage containers (polyethylene terephthalate), and corrugated containers. Communities are constantly adding to the list of MSW components to be collected for recycling. Materials recovery facilities (MRFs) are plants in which recyclables are prepared for shipment to remanufacturing facilities. The design of MRFs depends on the extent to which recyclables are separated from refuse

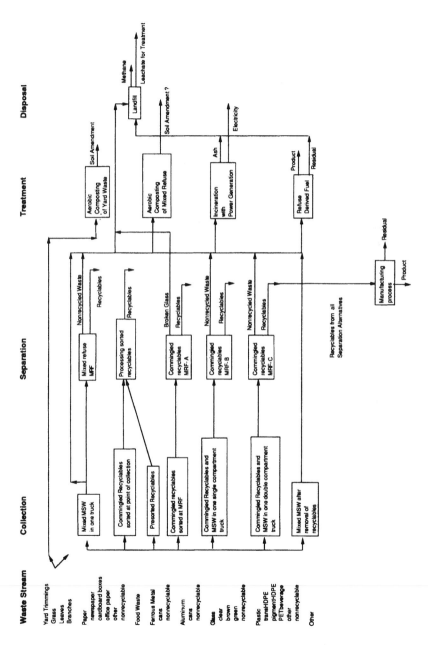

FIGURE 1 Alternatives for municipal solid waste management.

and each other prior to delivery to the MRF; several alternate collection/MRF strategies are presented in Figure 1.

Integrated solid waste management also may include combustion, an abiotic thermal treatment process that reduces the volume of waste requiring burial and offers the opportunity for energy recovery. Combustion is the major nonbiological treatment applied to MSW, and approximately 16% of the MSW generated in 1993 was oxidized by combustion. The typical combustion facility includes energy recovery as well as sophisticated equipment to control stack emissions. A second nonbiological process, referred to as *refuse-derived fuel* (RDF), also involves thermal oxidation. RDF is produced by separation of MSW into a product stream with a relatively high BTU value and a residual stream with a relatively low BTU value. The high BTU stream is then burned for energy recovery either in a dedicated combustion facility or in an industrial or utility boiler. There are many variations on the RDF theme, including the production of shredded refuse for direct combustion and the production of pellets for shipment over longer distances. Of course, only the organic fraction of MSW will be oxidized to carbon dioxide and water. The inorganics present in MSW will remain as either bottom ash or fly ash, which is the particulate material captured in the air pollution control equipment downstream of a combustion chamber. Ash resulting from MSW combustion typically is buried in a landfill.

Composting is a biological treatment option. It is most commonly applied to yard waste, although there are a number of facilities in the U.S. that compost MSW. The degree of pre- and post-treatment of the MSW will have a strong influence on potential uses of the finished compost product. Where composting is employed strictly for volume reduction prior to burial in a landfill, no pre- or post-processing is required. Conversely, if the material is to be used as a soil amendment, then extensive removal of noncompostable waste would be required.

The final process illustrated in Figure 1 is landfilling. A landfill is an ultimate disposal alternative, although decomposition of MSW in landfills does occur. As stated above, 62% of the MSW generated in the U.S. in 1993 was buried in landfills. Given that there are components of MSW which cannot be recycled or composted and that combustion is not the solid waste management alternative of choice for many communities, landfills will be an integral part of solid waste management for the foreseeable future.

As illustrated in Figure 1, there are a number of alternatives for MSW management, and they are highly interrelated. Decisions made with respect to the manner in which waste is collected and the types of waste that are recycled will affect the energy content of the MSW remaining for combustion, MRF design, and the mass and biodegradability of refuse remaining for burial. Integrated solid waste management requires consideration of upstream and downstream effects in the design of a solid waste management strategy.

1.4 BIOLOGICAL DECOMPOSITION

Depending on the method of disposal, MSW supports microbial communities which result in biological decomposition by aerobic or anaerobic processes. As our knowledge of these microbial communities increases, systems may be designed and managed to optimize MSW degradation and product formation. In this section, microbiological processes occurring in landfills, anaerobic digesters, and composting facilities are introduced. More detailed information is presented in Chapters 2, 3, and 4, which are devoted to these treatment and disposal alternatives.

1.4.1 Landfill Microbial Communities

The microbiology of landfill sites, including landfill leachate treatment and the soil cover, has been reviewed recently in Senior (1995). Our focus here will be on the microbiological transformations of MSW within landfills. A detailed review of landfill microbiology is found in Chapter 2. Earlier studies on landfill microbiology focused on aerobic rather than anaerobic microbial populations. The emphasis on aerobes was somewhat surprising because they play only a minor role in refuse decomposition and gas production in landfills. For example, Cook et al. (1967) isolated aerobic fungi, streptomycetes, and photosynthetic bacteria from a sanitary landfill, fresh household refuse, and seepage from the landfill. Public health concerns provided the incentive for several studies of landfill microbiology which focused on the isolation of indicator organisms such as coliforms and fecal streptococci from model and full-scale landfillls receiving both muncipal and hospital waste (Donnelly and Scarpino, 1984). In the 1980s, anaerobic microbiology in landfills became the subject of several research groups, particularly in the U.S. and the United Kingdom. Laboratory-scale model landfills or lysimeters (Figure 2) that simulate and accelerate anaerobic waste decomposition have greatly facilitated microbiological studies (Barlaz et al., 1989). Understanding the establishment and maintenance of anaerobic communities in landfills is critical to improving leachate quality, thereby reducing the risk of environmental impacts of landfills, such as groundwater contamination (Harper and Pohland, 1988).

Anaerobic consortia within landfills mediate the processes of polymer hydrolysis, fermentation to organic acids, and mineralization by methanogenesis (Barlaz et al., 1990; Senior and Balba, 1990). As described in Chapter 2, the chemical composition of MSW includes polymers such as cellulose, starch, protein, and lignin. Polymeric substrates in landfills are generally solid, high molecular weight compounds and, therefore, do not readily permeate microbial membranes. Thus, the hydrolysis of polymers to lower molecular weight substances is the first step in their biodegradation in landfills. Pavlostathis and Giraldo-Gomez (1991) have suggested that polymer hydrolysis can be the rate-limiting step for biodegradation of solid wastes in landfills. Extracellular

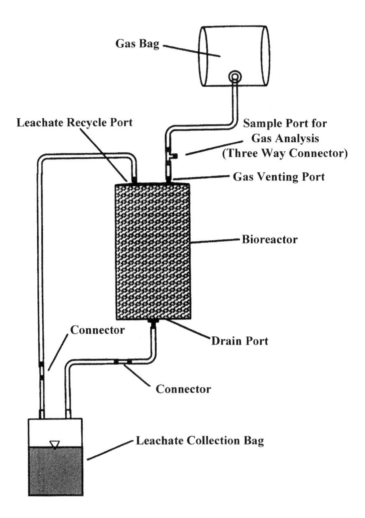

FIGURE 2   Laboratory-scale landfill lysimeter for studying decomposition of MSW.

enzymes must be produced by landfill microbes to hydrolyze polymers for subsequent biodegradation. Fermentative bacteria typically mediate polymer hydrolysis in anaerobic ecosystems (McInerney, 1988). Monomers resulting from polymer hydrolysis are then fermented to organic acids, which may be mineralized by methanogenic consortia in landfills (Barlaz et al., 1989). The degradation of organic acids to methane requires species of acetogenic bacteria together with H_2/CO_2 and acetate-utilizing species of methanogenic bacteria (Archer, 1984). Sulfate-reducing bacteria have been found to be abundant in several landfills and also may play a role in the mineralization of organic carbon in landfills (Suflita et al., 1992).

Barlaz et al. (1989) described four phases of refuse decomposition in landfills based on chemical and microbiological changes in laboratory reactors simulating landfill processes. In phase one, aerobic decomposition of refuse occurs while the oxygen entrained during refuse burial is rapidly depleted; nitrate also may be consumed by the process of denitrification in this phase. Once anaerobic conditions are established, phase two, the anaerobic acid phase, begins; fermentation of sugars results in the rapid accumulation of carboxylic acids and a concomitant decline in pH from about 7.5 to 5.7. Phase three, the accelerated methane production phase, is characterized by a rapid increase in the rate of methane production; consumption of carboxylic acids results in an increase of refuse pH. The fourth and final phase of refuse decomposition is the decelerated methane production phase. Here, the methane production rate is dependent on polymer hydrolysis as the accumulation of soluble fermentation intermediates has been depleted.

The major factors limiting methane production from landfills are moisture and pH. MSW is typically about 20% moisture by wet weight when buried. While this is sufficient for some biological activity, it is suboptimal. A landfill is typically operated to minimize moisture infiltration and leachate production. Thus, refuse may remain dry unless there is substantial infiltration of precipitation. As discussed above, refuse pH is likely to decrease due to an imbalance between fermentative, acetogenic, and methanogenic activities.

To enhance the rates of anaerobic decomposition, methane production, and leachate stabilization, landfills could be operated as bioreactors (Townsend et al., 1995). This would require that precipitation be allowed to infiltrate the landfill. The increased infiltration would result in a higher moisture content in the landfill and increased leachate production. To enhance methane production, this leachate could be recycled back over the top of the refuse. Prior to recycling, the leachate could be neutralized to facilitate transition of the refuse from the anaerobic acid phase to the accelerated methane production phase of decomposition. The concepts of leachate recycling and neutralization have been demonstrated in laboratory-scale reactors (Pohland, 1975, 1980; Barlaz et al., 1987). There are few published data on the results of leachate recycling and neutralization in field-scale landfills, either with respect to methane production rate or to effects on the microbial community. Nonetheless, operating landfills as bioreactors is widely expected to enhance methane production rates and total yields.

Landfilled refuse provides a habitat for morphologically and metabolically diverse microbes in landfill microbial ecosystems (Archer and Peck, 1989). Bacteria colonize refuse surfaces (Figure 3) and may be found free-living in landfill leachates. Sleat et al. (1987) pointed out that for biological stabilization of landfills, "the numbers and activities of the key groups of microbes must be adequately high and fully integrated, and the physical and chemical environment of the landfill must be conducive to their effective growth and activity." The authors identified seven key groups of microbes in the Aveley Landfill, Essex, U.K.: amylolytic, proteolytic, cellulolytic, and hemicellulolytic bacteria; hydrogen-oxidizing methanogens; acetoclastic methanogens; and sulfate-reducing bacteria.

FIGURE 3   The ciliate *Metopus palaeformis* from landfilled refuse. The protozoan is filled with symbiotic methanogens which are autofluorescent. The ciliate is about 100 μm in length. (From Finlay, B.J., and T. Fenchel. 1991. *FEMS Microbiol. Ecol.* 85:169-180. With permission.)

To date, few bacteria have been isolated and characterized from full-scale landfills. While cellulose is the predominant substrate in most landfills, typically representing about 40% of the inherently biodegradable substrate, few reports have been made of the isolation of anaerobic cellulolytic bacteria from refuse (Bagnara et al., 1985; Westlake, 1989). Westlake (1989) cultured five cellulolytic anaerobic isolates from several landfill sites in the U.K.; maximum rates of cellulose hydrolysis by the isolates were found at a pH of 4 to 5. The activity was cell associated, like other reported bacterial cellulases. More recently, Westlake et al. (1995) isolated an aerotolerant, cellulolytic *Clostridium* sp. and three obligately anaerobic cellulolytic *Eubacterium* spp. with temperature optima of 37°C or higher from landfilled refuse. This suggests that landfilled refuse supports a diversity of cellulose-degrading microorganisms.

Fielding et al. (1988) isolated the methanogens *Methanobacterium formicicuma* and *Methanobacterium bryantii* from several landfills in the U.K. Peck and Archer (1989) have used the quantification of the coenzyme F420 to estimate the methanogenic biomass in landfill samples. An anaerobic protozoon (the ciliate *Metopus paleaeformis*) with symbiotic methanogens was recently identified in landfilled refuse (Figure 4; Finlay and Fenchel, 1991). Approximately 300 bromoethanesulfonate-sensitive methanogenic symbionts were detected by autofluorescence in each cell. It is assumed that, since the

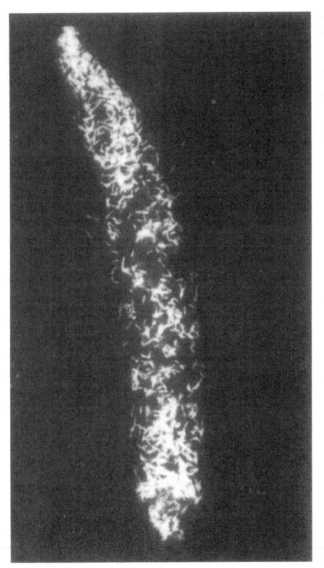

FIGURE 4 Scanning electron micrograph showing morphologicaly diverse bacteria associated with the surface of a grass blade recovered from a landfill. (From Suflita, J.M. et al. 1992. *Environ. Sci. Technol.* 26:1486–1495. With permission of the American Chemical Society.)

hydrogenosomes contained an hydrogenase, *M. paleaeformis* disposes of reducing equivalents as hydrogen gas, which is then captured and oxidized by the methanogens. Availability of water probably restricts growth of protozoa in landfills, and ciliates may remain encysted at drier sites. The presence of protozoa is important because it suggests that a microbial food web may occur in landfilled refuse.

Leachates result as water percolates through landfills, and organic and inorganic constituents are dissolved. Characteristics of leachate are highly variable depending on refuse composition, particle size, soil cover, moisture, content, water application, landfill design and operation, degree of compaction, operation, and age (Pohland and Harper, 1987; Lisk, 1991). Christensen et al. (1994) reported ranges of general landfill leachate parameters from the literature (Table 3). In the initial acid phase, pH is low (4.5 to 5.5) and easily degradable organic compounds such as volatile fatty acids contribute to a high biological oxygen demand (BOD). In the methanogenic phase, pH increases (7.5 to 9) and the BOD decreases sharply. Leachates from poorly managed landfills may inadvertently enter neighboring aquifers, resulting in potential contamination with toxic chemicals. Landfill leachates are typically highly reduced and enriched in carbon relative to the aquifers, and thus they can alter the natural microbiota as well as many key physical-chemical processes such as redox chemistry and sorption (Beeman and Suflita, 1987; Christensen et al., 1994). The high biodegradative potential of landfill leachates suggests the presence of metabolically diverse microorganisms that are either free-living or attached to small particulates.

1.4.2 Anaerobic Digestion of MSW

Anaerobic digestion of the biodegradable organic fraction of MSW under controlled conditions allows for the acceleration and optimization of degradation processes that occur in landfills. While anaerobic digestion of waste sludges with energy recovery has been an economically viable source of energy production for over 50 years, application to MSW has not found widespread commercial acceptance (Stenstrom et al., 1983). Anaerobic digestion of MSW has developed more rapidly in Europe during the last decade; however, it is not as common in the U.S. (see Chapter 3).

The biodegradable organic fraction of MSW includes food, yard, and paper wastes that are high in cellulose and can serve as a substrate for anaerobic digestion. Pathways for biodegradation of MSW in anaerobic digesters are similar to those in landfills and other anaerobic environments. Polymer hydrolysis is followed by fermentation to organic acids and ultimate biodegradation to methane and CO_2. The final products of anaerobic digestion are methane, carbon, and a stable, humus-like material; the latter often requires subsequent aerobic composting prior to use as a soil amendment (Vallini et al., 1993). Feedstock biodegradability is critical to the economic viability of this approach

TABLE 3   Landfill Leachate Composition in Acid and Methanogenic Phases[a]

Parameter	Acid Phase Average	Acid Phase Range	Methanogenic Phase Average	Methanogenic Phase Range	Average for Both Phases[b]
pH	6.1	4.5–7.5	8	7.5–9	
BOD[c]	13,000	4000–40,000	180	20–550	
COD[d]	22,000	6000–60,000	3000	500–4500	
Sulfate	500	70–1750	80	10–420	
Calcium	1200	10–2500	60	20–600	
Magnesium	470	50–1150	180	40–350	
Iron	780	20–2100	15	3–280	
Manganese	25	0.3–65	0.7	0.03–45	
Ammonia-N	—	—	—	—	741
Chloride	—	—	—	—	2120
Potassium	—	—	—	—	1085
Sodium	—	—	—	—	1340
Total phosphorus	—	—	—	—	6
Cadmium	—	—	—	—	0.005
Chromium	—	—	—	—	0.28
Cobalt	—	—	—	—	0.05
Copper	—	—	—	—	0.065
Lead	—	—	—	—	0.09
Nickel	—	—	—	—	0.17
Zinc	5	0.1–120	0.6	0.03–4	

[a] All values are in mg/l except pH and BOD/COD.

[b] Average values for parameters with no observed differences between acid and methanogenic phases.

[c] Five-day biological oxygen demand.

[d] Chemical oxygen demand.

Source: Christensen, T.H. et al. 1994. *CRC Crit. Rev. Environ. Sci. Technol.* 24: 119-202. With permission.

(Kayhanian, 1995). To increase biogas yield and the quality of humic-like residue, it is necessary to minimize nondegradable compounds by careful presorting of wastes (Cecchi et al., 1992). Anaerobic digestion of the organic fraction of MSW can proceed at high solids concentrations (>20% solids) in the feed (Van Meenen et al., 1988; Van Meenen and Verstraete, 1988). A complete description of the anaerobic digestion process and common digester designs is presented in Chapter 3.

Since cellulosic materials are the major biodegradable component of MSW, cellulose hydrolysis is a key step in biodegradation. Hydrolytic enzymes in anaerobic digesters were studied by Adney et al. (1989). Detergent extraction was required for the recovery of cellulase, beta-glucosidase, alpha-glucosidase, protease, and alpha-amylase from sludge particulates. Their results suggested that most hydrolytic enzymes are strongly cell associated. Benoit et al. (1992)

isolated ten strains of obligately anaerobic, cellulolytic bacteria from a MSW digester used for biogas production. The optimum growth temperatures of the strains ranged between 34 and 40°C, and the optimum pH for growth was 7.5 to 7.8 for most of the strains. The gram positive, spore-forming rods were identified as clostridia based on their biochemical characteristics. Most strains degraded 15 to 35% of shredded newspaper and magazine paper within 15 days. Further characterization of these isolates by molecular methods was performed by Benoit and coworkers (Cailliez et al., 1992). Hybridization experiments were conducted with cloned gene fragments from *Clostridium cellulolyticum*. Two endoglucanase genes hybridized with most strains, suggesting homology and widespread distribution of those genes. None of the strains hybridized with *nif* genes, however, indicating that the digester clostridia did not appear to fix nitrogen. On the basis of the hybridization and biochemical tests, the authors concluded that these strains were different, not only from *C. celluolyticum*, but from other clostridia described in the literature.

Temperature has been shown to have a profound effect on the rates of anaerobic biodegradation of MSW. To optimize the process of anaerobic digestion, biodegradation at mesophilic (37°C) and thermophilic (55°C) temperatures was compared by Cecchi et al. (1991). The semi-dry thermophilic process had a gas production rate two to three times the mesophilic process. Cecchi et al. (1992) found that biodegradation rates in digesters exposed to ambient environmental temperatures were greater in the summer (>18°C) than the winter (2°C).

Reactor configurations vary in complexity, ranging from the continuously stirred tank reactor (CSTR) used in fermentation technologies to multiple-stage batch reactors (see Chapter 3). In CSTR, the total solids concentration typically ranges from 3 to 8% to achieve satisfactory mixing. Two-phase digestion, suggested by Ghosh et al. (1975), involves hydrolysis and acidification in the first reactor, with methanogenesis in a second reactor. High solids fermentation (>20% total solids) is also referred to as *"dry" anaerobic digestion* or *solid state fermentation*. De Baere et al. (1985) described a solid state fermentation at 30 to 35% total solids followed by a post-digestion process in which the digested residue is dewatered and the liquid fraction is recycled to inoculate fresh substrate. This process is sometimes referred to as *anaerobic composting;* however, this is a misnomer because composting is by definition an oxygen-consuming process. Chynoweth et al. (1991) described a three-stage, sequential batch reactor for anaerobic digestion that utilizes leachate to provide organisms, moisture, and nutrients to enhance biodegradation of MSW to methane. In stage 1, the biodegradable fraction is coarsely shredded, moistened, and placed in a reactor inoculated with recycled leachate; most of the methane is produced in this stage. In stage 2, the fermentation is active and balanced, and methane production levels off. Stage 3 allows for the completion of bioconversion of particulates and serves as an inoculum for stage 1.

1.4.3 Composting of MSW

Composting is the microbial degradation of organic solid material that involves aerobic respiration and passing through a thermophilic stage (Finstein and Morris, 1975). A detailed review of compost microbiology is presented in Chapter 4. Composting involves physical, chemical, and microbiological changes to the composting mass. These include self-heating of the mass, an increase in microbial activity and biomass, and a decrease in volatile solids and C:N ratios. The products of the aerobic microbiological transformation of refuse are CO_2 and a humus-like material which is comprised primarily of stable, lignocellulosic compounds. Microbial community diversity is desirable in composting because it promotes community stability. Such diversity protects the community from perturbations and increases metabolic versatility, allowing the utilization of the widest range of substrates.

In the U.S., composting of MSW currently accounts for about 1% of the total MSW disposed (USEPA, 1994a). Unlike landfilling, composting has a product orientation that encourages process optimization. The composting process results in a reduction in solid waste mass, a reduction in pathogenic organisms, and the recovery of a potentially marketable soil amendment which is rich in nitrogen (Keeling et al., 1994). He et al. (1995) concluded that MSW-derived compost is likely to improve soil properties because of its low bulk density, high water-holding capacity, slightly alkaline pH, and high organic matter content.

The composting process can be divided into four phases based on ambient temperature: a mesophilic phase (phase I), a thermophilic phase (phase II), a cooling phase (phase III), and a maturation phase (phase IV) (Gray et al., 1971; Finstein and Morris, 1975). In phase I, the compost mass is at ambient temperature and may be slightly acidic. As the indigenous organisms multiply, the temperature rises rapidly, and the compost material acts as an insulator. Production of simple organic acids during this phase can cause a slight drop in pH. In phase II, the temperature exceeds 40°C, the mesophiles decline, and degradation is dominated by the thermophiles. The pH increases, and ammonia may be liberated following protein deamination. In this thermophilic phase, fungi secrete extracellular enzymes which break down polymers such as cellulose and other complex carbohydrates. As readily degradable substrates decline, heat loss exceeds metabolic heat generation and phase III, the cooling phase, is initiated. At temperatures below 40°C, mesophilic organisms recommence activity and the pH drops slightly. Nitrifying bacteria, which had been inhibited at the higher temperatures, begin to convert ammonia to nitrate.

Phase IV, compost maturation, is critical to its agronomic use. An immature compost can introduce phytotoxic materials to the soils, such as ammonia or volatile fatty acids, and it can result in a decrease in the Eh of the amended soil (Iglesias-Jimenez and Perez-Garcia, 1992). Iglesias-Jimenez and Perez-Garcia (1992) proposed that guidelines for mature compost should include

(1) a stabilization of the temperature curve, (2) a water soluble C:N ratio lower than 6, (3) a ratio of CEC to total organic carbon higher than 1.9, and (4) a ratio of humic acids to fulvic acids that is higher than 1.9.

Temperature, aeration, moisture, pH, refuse type, particle size, and carbon to nitrogen ratio are some of the factors affecting composting. Temperature is the most critical parameter in the composting process (Suler and Finstein, 1977). Microbial activity and diversity decrease above 55 to 60°C (McKinley and Vestal, 1985a,b; Strom, 1985a,b) seriously limiting the rate of decomposition of organic matter. While temperatures above 60°C inhibit the composting process, only very low concentrations of pathogens (both animal and plant) can be tolerated, and the regrowth of such pathogens must be discouraged in a soil amendment. A discussion of the detection of pathogenic microbes during MSW composting is presented in Chapter 5. Heat generated during the thermophilic phase of composting is critical for reducing the number of pathogens. Pathogenic organisms are destroyed, provided the composting temperature reaches 60 to 70°C for a minimum of 3 consecutive days.

Moreover, MSW composting may be a source of nuisance odors from volatile emissions. Kissel et al. (1992) reported that emissions from MSW composting included inorganic and organic compounds of sulfur and nitrogen, low molecular weight aliphatics, terpenes, carbonyls, and alcohols. Volatile organic chemicals (VOC) typically are emitted in the early stages of MSW composting and may include both xenobiotic chemicals and compounds of natural origin such as terpenes (Eitzer, 1995). Derikx et al. (1990a) found hydrogen sulfide, carbonyl sulfide, carbon disulfide, dimethyl sulfide, dimethyl disulfide, and dimethyltrisulfide to be the most common odor-forming compounds during composting. The souces of sulfur were unclear but may include gypsum or sulfur-containing amino acids. The optimal temperature for the production of hydrogen sulfide, carbonyl sulfide, methanethiol, and dimethyl sulfide paralleled the optimal temperature for biological activity (50 to 56°C), suggesting that the sulfur-containing compounds are of biological origin (Derikx et al., 1990b). The development of anaerobic microenvironments in poorly mixed systems may allow conditions favoring the microbiological formation of volatile sulfur compounds. Increased aeration greatly reduced the production of these compounds (Derikx et al., 1990b).

1.5 SOCIOECONOMIC ISSUES

If improperly disposed, solid waste can pose threats to human health and the environment via gaseous emissions, leachate, combustor ash, and litter. In both terrestrial and aquatic habitats, accumulation can be prevented only if rates of decomposition exceed rates of input. This is especially problematic in marine environments where recalcitrant litter can result in the deaths of marine mammals and birds.

1.5.1 Public Health Concerns

The primary human health concern is the potential for contamination of our drinking water with pathogens and toxic chemicals. As recently as 1986, there were an estimated 6000 operating municipal landfills in addition to thousands of closed facilities (USEPA, 1988). Most of these sites were constructed without an engineered liner or leachate collection system, and nearly 50% of these sites were within 1.6 km of a drinking water well. Thus, the potential for groundwater contamination from pathogens and xenobiotics from MSW landfills is of significant concern. Information on design strategies employed to contain refuse and leachate within a landfill, thereby separating it from the environment, is presented in Chapter 2. Unfortunately, landfills designed to contain buried refuse are a relatively recent phenomenon, and historically refuse was simply placed in a large excavation with little consideration for the underlying groundwater. On the east coast of the U.S., many older landfills were built on marshland; thus, certain near-shore marine habitats have been negatively impacted by leaking landfills (Johnston et al., 1994). Biomagnification in marine food chains is a serious problem, particularly with fat-soluble xenobiotic chemicals such as PCBs and with metals. The primary public health concern with microbial pathogens is the potential for transport and sustained viability during solid waste disposal. Our discussion here will primarily address the fate of pathogens and xenobiotics in landfills because this is still by far the most common disposal option for MSW.

The potential for groundwater contamination with human pathogens from refuse has long been of concern. A detailed review of microbial pathogens in MSW can be found in Chapter 5. The primary pathogens of concern are viruses such as enteroviruses, hepatitis A, rotaviruses, and the protozoa *Giardia* and *Cryptospiridium*. Sources include sewage sludge and septage which are sometimes co-disposed with MSW, and pet and human (from disposable diapers) excreta (Pahren, 1987). Cook et al. (1967) isolated fecal indicators *Escherichia coli* and *Streptococcus faecalis* from fresh and partially decomposed refuse and seepage. Donnelly and Scarpino (1984) isolated indicator microorganisms from full-scale and laboratory landfills to determine their survival under landfill conditions. Total and fecal coliforms decreased rapidly within 13 weeks to <20 per 100 ml leachate, and fecal streptococci disappeared after 2 years. Sobsey (1978) examined 22 leachate samples from landfills in the U.S. and Canada for enteric viruses. He concluded that properly operated landfills do not constitute a health hazard from enteric viruses because of the low concentration of viruses in raw leachates, thermal inactiviation, removal in soil during percolation, and dilution in groundwater and surface waters. More recently, nucleic acid probes as well as tissue culture have been used to study the persistence of enteric viruses in landfilled disposable diapers (Huber et al., 1994). It is important to note that, unlike cell culture, nucleic acid probes detect the presence of viral genomes, not the viability of viruses. While Huber et al. (1994) found that poliovirus RNA was present in some diapers, the polioviruses were not viable by cell culture assays.

Leachate from older landfills frequently has been the cause of groundwater contamination. Christensen et al. (1994) have reviewed the literature on the attenuation of organic pollutants in landfill leachate-contaminated aquifers and provide a summary of organic chemicals measured in contaminated groundwater (see Table 3). Because landfill leachate typically has relatively high levels of organic carbon, the naturally present dissolved oxygen in an aquifer is rapidly depleted and anaerobic conditions prevail. Leachate composition and aquifer biogeochemistry will determine the dominant oxidation-reduction processes. Terminal electron acceptors that may be important include nitrate or sulfate, if these compounds are present in either the aquifer or the leachate; Fe (III) or Mn (IV), if the aquifer contains these minerals in its sediment; or carbon dioxide, if methanogenensis is the dominant degradative process.

The impact of a leachate-contaminated aquifer on human health and the environment is a function of a number of site-specific parameters, including the rate of attenuation within the aquifer, the proximity of drinking water wells, and the potential discharge of the leachate to surface water. In the case of contaminants which are more readily degradable under aerobic conditions, leachate discharge to surface water may enhance biodegradation. In the case of volatile compounds, a discharge to surface water will likely result in their release to the atmosphere.

1.5.2 Contribution to Greenhouse Gases

Methane and carbon dioxide are greenhouse gases that contribute to global climate change (Rogers and Whitman, 1991). Methane is estimated to be approximately 20 times more damaging than carbon dioxide on a volume basis (USEPA, 1994b). Consequently, although landfill gas contains approximately equal proportions of methane and carbon dioxide, methane is more significant with respect to atmospheric climate change. Based on recent estimates of landfill (USEPA, 1995) and total (USEPA, 1994b) anthropogenic methane emissions, landfills are responsible for approximately 8% of total anthropogenic methane releases (Figure 5). Alternatives for the management of landfill gas include (1) allowing the gas to vent naturally to the atmosphere through the landfill cover, (2) venting the gas to the atmosphere through a series of wells installed through the landfill cover and into the refuse, (3) recovering the gas and routing it to a flare for conversion of the methane to carbon dioxide, or (4) recovering the gas for its energy value. In 1992, methane was recovered in commercial quantities at 119 landfills in the U.S. and Canada (Thorneloe and Pacey, 1994). Typically, this methane is converted to electricity, although in some cases it is piped directly into industrial boilers or cleaned up to natural gas pipeline standards. Under revisions to the Clean Air Act, the USEPA requires the largest landfills in the U.S. to capture their landfill gas and either convert the methane to carbon dioxide using a flare or recover the energy value of the methane.

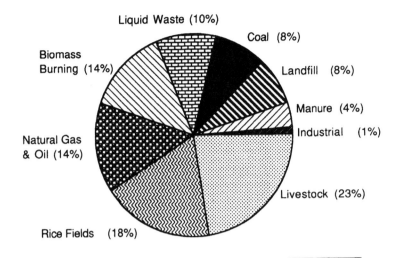

Total Anthropogenic Methane: 354 (277 - 477) Tg/yr

Landfills: 30 (19 - 40) Tg/yr

FIGURE 5 Relative contributions of anthropogenic sources of global methane emissions. (From USEPA. 1994b. International Anthropogenic Methane Emissions: Estimates for 1990. EPA 230-R-93-010. U.S. Environmental Protection Agency, Office of Policy Planning and Evaluation, Washington, D.C.)

1.5.3 Biodegradability of Polymeric Materials in MSW

Public concern over the disposal and lack of degradability of plastics has escalated in recent years. Most plastics currently in use are polyolefin derived, such as polyethylene and polypropylene, and they are designed not to degrade. The development of biodegradable plastics has been proposed as a solution for the problem of waste plastics. The consumer has been swamped with claims of biodegradability of everything from diapers to trash bags.

Plastics pose some difficult problems for microbially mediated degradation because they are insoluble and hydrophobic. Polymeric substrates in landfills are generally solid and of a high molecular weight and, therefore, do not readily permeate microbial membranes. Extracellular enzymes must be produced to hydrolyze these polymers. One reason that polyethylene is not biodegradable is that it lacks a point of hydrolytic attack other than the terminus of the carbon chain, which is located in the bulk of the hydrophobic matrix. To degrade a polymer, microbial enzymes need access to polymer sites, and polymers must have chemical groups susceptible to enzymatic attack.

What does biodegradability of plastics mean in this context, considering the slow rates of biodegradation typical of landfills, currently the primary disposal option? Palmisano and Pettigrew (1992) proposed that an understanding of the biodegradability of plastics must include (1) knowledge of the inherent biodegradability of the polymer and its breakdown products based on molecular structure and physical form; (2) determination of rates of biodegradation in relevant disposal matrices, such as landfills and composters; and (3) a comparison of rates of biodegradation to rates of input to assess potential accumulation over time. Candidates to replace polyethylene and polypropylene include

1. Polyethylene blends with inherently biodegradable polymers, such as starch or polycaprolactone
2. Synthetic polymers with a weak link or bonds that are susceptible to hydrolysis, photolysis, or oxidation
3. Bacterially derived polyesters such as the polyhydroxyalkanoates
4. Polyolefins amended with pro-oxidants which are often mixtures of transition metals and lipids that produce free radicals in the presence of heat or light
5. Cellulose- or starch-based materials

In recent years, significant progress has been made in the development of biodegradability testing for plastics and other polymers under realistic disposal conditions. Tiered testing approaches for polymer biodegradation are discussed in Chapter 6. Briefly, there are three tiers of testing which involve increasing complexity and cost. The first tier consists of screening level tests wherein the inherent biodegradability of a polymer is examined under conditions that are conducive to biodegradation. Examples include the growth of pure cultures or mixed cultures, enzyme assays, or carbon dioxide production to assay biodegradation. The second tier includes confirmation level tests that examine biodegradability under controlled conditions that attempt to simulate real world conditions. Confirmation level tests may utilize isotopically labeled materials (when available) which are critical for determining pathways of polymer degradation. They also may include the use of microcosms or mesocosms designed to simulate the disposal environment. The third and ultimate tier is field testing wherein demonstration of *in situ* biodegradation is used to determine whether the parent material or intermediates persist in the environment.

1.6 DECISION-MAKING TOOLS FOR SOLID WASTE MANAGEMENT

Product life-cycle analysis attracted attention in the late 1980s as a method for identifying the resource requirements as well as the environmental impacts associated with each stage in the life of a product, from "cradle-to-grave". Portney (1993/1994) described three phases of product life cycle assessment. The inventory phase involves determining the environmental loading, such as the amounts of energy and raw materials consumed, and the amount of pollution, including solid waste, generated at each stage. The impact phase translates this inventory into damage done to human and environmental health. The improvement phase aims to decrease the impact or adverse effects of the product life cycle by identifying processes that contributed to them. Portney sounded a cautionary note in his essay: It is difficult to delineate and quantify all aspects of most product life cycles, and the results can sometimes be misleading to the consumer.

Recently, the USEPA and the U.S. Department of Energy (USDOE) jointly initiated a project to perform a life-cycle assessment of solid waste management alternatives (Thorneloe et al., 1995). The objective of the project is to develop a decision support system that will allow solid waste management planners and policy makers to evaluate alternate solid waste management strategies in consideration of a number of variables, henceforth referred to as *life-cycle inventory* (LCI) parameters. LCI parameters include gaseous emissions such as CO_2, NO_x, and SO_x; water contaminants such as biological oxygen demand; solid waste generation; and energy consumption. The cost of each management strategy can be evaluated also in the decision support system.

The LCI analysis will include all aspects of a solid waste management strategy, including waste collection and transport, separation, combustion, composting, and burial. In cases where wastes are recycled, the analysis will include a comparison of the resources and emissions associated with production of a new product from either recycled or virgin materials. When complete, the decision support system will enable a user to compare, for example, CO_2 emissions associated with recycling, composting, burning, or burying old newspapers in the context of MSW management. A simplified version of this decision support system, which incorporates economics only, has been completed for the processes included in Figure 1 (Barlaz et al., 1995; Ranjithan et al., 1995).

Ultimately, information on life-cycle parameters for solid waste management alternatives should form a component of the analysis used by designers of new products and packaging materials. Downstream environmental release data should be incorporated in the evaluation of alternative material and product designs. While not strictly a problem for microbiologists, understanding the biological fate of materials plays a significant role in life-cycle analyses as exemplified by the importance of landfills to anthropogenic methane emissions.

1.7 RESEARCH GAPS AND OPPORTUNITIES

1.7.1 Community Analysis by Molecular and Biochemical Methods

In the past, our inability to culture most microorganisms in nature has severely restricted our ability to understand the community structure and function in many environments, including decomposing solid waste. The modern tools of molecular biology and biochemistry provide a means for characterizing microbial communities in MSW without the need to culture microorganisms. The 16S rRNA probes could be useful in identifying the relative proportion of microbes in the kingdoms of Archaea, Eubacteria, and Eukaryota. Community analysis methods that "fingerprint" communities at various phases of decomposition in landfills and composting would be especially valuable. Recently, Malik et al. (1994) extracted DNA from a composter containing a synthetic solid waste feed. A randomly primed polymerase chain reaction (PCR) was used to create a DNA fingerprint for composting microbial communities. Denaturing gradient gel electrophoresis (DGGE) is emerging as a powerful tool in characterizing microbial communities (Muyzer et al., 1993). Analysis of fatty acids associated with membrane-bound phospholipids in microbes also has proven to be a reliable and comprehensive method of community analysis (Vestal and White, 1989).

1.7.2 The Role of Bactivory and Lysis in Refuse-Associated Microbial Communities

Little is known about the turnover of the relatively high densities of microorganisms associated with decomposing MSW. In soils, freshwater, and marine environments, bacterial predators play an important role in determining both biomass and activity of bacterial populations. The identification of several types of ciliates in landfills (Finlay and Fenchel, 1991) suggests that bactivory may play a role in maintaining bacterial population densities in landfills. The same may be true in anaerobic digesters or composters.

Recent reports have demonstrated that, in marine environments, lysis of bacteria by viruses (bacteriophage) also may be important in controlling bacterial population densities (Furhman and Noble, 1995). To date, only viruses that are human and plant pathogens have been examined in decomposing refuse. Lysis by bacteriophage, however, may have a profound impact on the stability of bacterial populations in landfills, composters, and anaerobic digesters.

1.7.3 Surface Analysis of Microbial Communities

New methods of microscopy coupled to image analysis are allowing microbial ecologists to examine the interaction of microbes with surfaces. Microorganisms can be digitally imaged using fluorescent molecular probes, confocal laser microscopy, and computer image analyses (Caldwell et al., 1992). Confocal laser microscopy has emerged as a very powerful technique which allows one to analyze and quantify a three-dimensional microbial assemblage in a nondestructive manner. While this method has proven useful in the study of biofilms and bioaggregates, it also could be applied to microbes colonizing refuse surfaces.

Another new microscopic method is the use of the environmental scanning electron microscope (SEM). This instrument differs from conventional SEM in that the specimen can remain hydrated. The dehydration and critical point drying required for traditional electron microscopy undoubtedly distorts the architecture of microbial communities. The environmental SEM has been applied successfully by Little and coworkers (McNeil et al., 1991) to examine microbially induced corrosion and could provide information on the association of microbes with refuse surfaces in the presence of moisture.

1.7.4 Designing Polymers for Hydrolysis

As a society, we continue to search for polymeric materials that rapidly biodegrade to benign products. Weak-link technologies involve inserting potentially hydrolyzable bonds into polymers. For this strategy to be effective, however, we must develop an understanding of the array of hydrolytic enzymes that are elaborated by microbial communities in landfills, composters, and anaerobic digesters. Environmental factors affecting the regulation and activity of hydrolytic enzymes by bacterial populations need to be more closely examined. This information can be used to design materials which are readily biodegraded under typical conditions found in solid waste disposal.

1.8 CONCLUSIONS

Decomposing refuse provides both habitat and substrate for the growth of diverse microbial communities under anaerobic or aerobic conditions. Engineered systems such as landfills, digesters, and composting units are microbial ecosystems which require further study. A thorough understanding of the factors affecting refuse decomposition by microbiota would allow for the optimization of these processes. As the tonnage of disposed refuse continues to rise, the knowledge of safe, rapid means of biodegradation of the organic fraction of MSW is needed. Volatile emissions including greenhouse gases, transport of potential toxics in groundwater from leachates, and transport and viability of microbial pathogens will continue to be important societal issues. Our expanding knowledge of the microbiology of solid waste will be critical to dealing with these problems.

REFERENCES

Adney, W.S., C.J. Rivard, K. Grohmann, and M.E. Himmel. 1989. Detection of extracellular hydrolytic enzymes in the anaerobic digestion of municipal solid waste. *Biotech. Appl. Biochem.* 11:387–400.

Amann, R.I., W. Ludwig, and K.-H. Schleifer. 1995. Phylogenetic identification and in situ detection of individual microbial cells without cultivation. *Microbiol. Rev.,* 59:143–169.

Archer, D.B. 1984. Detection and quantitation of methanogens by enzyme-linked immunosorbent assay. *Appl. Environ. Microbiol.* 48:797–801.

Archer, D.B., and M.W. Peck. 1989. The microbiology of methane production in landfills, in M.S. da Costa, J.C. Duarte, and R.A.D. Williams (Ed.), *Microbiology of Extreme Environments and its Potential for Biotechnology.* Elsevier Applied Science, New York, pp. 187–204.

Bagnara, C., R. Toci, C. Gaudin, and J.P. Belaich. 1985. Isolation and characterization of a cellulolytic microorganism *Cellulomonas fermentans* sp. nov. *Int. J. Syst. Bacteriol.* 35:502–507.

Barlaz, M.A., M.W. Milke, and R.K. Ham. 1987. Gas production parameters in sanitary landfill simulators. *Waste Manage. Res.* 5:27–39.

Barlaz, M.A., E.D. Brill, Jr., A. Kaneko, S.R. Nishtala, HR. Piechottka, and S. Ranjithan. 1995. Integrated solid waste management: 1. Mathematical modeling. ASCE 2nd Congress on Computing in Civil Engineering, Atlanta, June 5–7, 1995, American Society of Chemical Engineers, New York.

Barlaz, M.A., D.M. Schaefer, and R.K. Ham. 1989. Bacterial population development and chemical characteristics of refuse decomposition in a simulated sanitary landfill. *Appl. Environ. Microbiol.* 55:55–65.

Barlaz, M.A., R.L. Ham, and D.M. Schaefer. 1990. Methane production from municipal refuse: a review of enhancement techniques and microbial dynamics. *Crit. Rev. Environ. Control* 19:557–584.

Beeman, R.E., and J.M. Suflita. 1987. Microbial ecology of a shallow unconfined ground water aquifer polluted by municipal landfill leachate. *Microbiol. Ecol.* 14:39.

Benoit, L., C. Cailliez, E. Petitdemange, and J. Gitton. 1992. Isolation of cellulolytic mesophilic clostrida from a municipal solid waste digester. *Microbiol. Ecol.* 23:117–125.

Bookter, T.J., and R.K. Ham. 1982. Stabilization of solid waste in landfills. *J. Environ. Eng.* 108:1089–1100.

Cailliez, C., L. Benoit, J.P. Thirion, and H. Petitdemange. 1992. Characterization of 10 mesophilic cellulolytic clostrida isolated from a municipal solid waste digester. *Curr. Microbiol.* 25:105–112.

Caldwell, D.E., D.R. Korber, and J.R. Lawrence. 1992. Confocal laser microscopy and digital image analysis in microbial ecology, in K.C. Marshall (Ed.), *Advances in Microbial Ecology.* Vol. 12. Plenum Press, New York, pp. 1–67

Cecchi, C., A. Marcomini, P. Pavan, G. Fazzini, and J. Mata-Alvarez. 1990. Mesophilic digestion of the organic fraction of refuse: peformance and kinetic study. *Waste Manage. Res.* 8:33–44.

Cecchi, F., P. Pavan, J. Mata-Alvarez, A. Bassetti, and C. Cozzolino. 1991. Anaerobic digestion of municipal solid waste: thermophilic vs. mesophilic performance at high solids. *Waste Manage. Res.* 9:305–315.

Cecchi, F., J. Mata-Alvarez, P. Pavan, G. Vallini, and F. De Poli. 1992. Seasonal effects on anaerobic digestion of the source sorted organic fraction of municipal solid waste. *Waste Manage. Res.* 10:435–443.

Christensen, T.H., P. Kjeldsen, H.J. Albrechtsen, G. Heron, P.H. Nielsen, L.B. Poul, and P.E. Holm. 1994. Attenuation of landfill leachate pollutants in aquifers. *CRC Crit. Rev. Environ. Sci. Technol.* 24:119–202.

Chynoweth, D.P., G. Bosch, J.F.K. Earle, R. Legrand, and K. Liu. 1991. A novel process for anaerobic composting of municipal solid waste. *Appl. Biochem. Biotechnol.* 28/29:421–432.

Cook, H.A., D.L. Cromwell, and H.A. Wilson. 1967. Microorganisms in household refuse and seepage water from sanitary landfills. *Proc. W. Va. Acad. Sci.* 39:107–114.

De Baere, L., O. Verdonck, and W. Verstraete. 1985. High rate dry anaerobic composting process for the organic fraction of solid wastes, in *Proceedings of the Biotechnology and Bioengineering Symposium No. 15*, John Wiley & Sons, New York, pp. 321–330.

Derikx, P.J.L., H.J.M. Op den Camp, C. van der Drift, L.J.L.D. van Griensven, and G.D. Vogels. 1990a. Odorous sulfur compound emitted during production of compost used as a substrate in mushroom cultivation. *Appl. Environ. Microbiol.* 56:176–180.

Derikx, P.J.L., H.J.M. Op den Camp, C. van der Drift, L.J.L.D. van Groemsven, and G.D. Vogels. 1990b. Biomass and biological activity during the production of compost used as a substrate in mushroom cultivation. *Appl. Environ. Microbiol.* 56:3029–3034.

Donnelly, J.A., and P.V. Scarpino. 1984. Isolation, Characterization and Identification of Microorganisms from Laboratory and Full-Scale Landfills. EPA Project Summary EPA600/S2-84-119:1-7. U.S. Environmental Protection Agency, Washington, D.C.

Eitzer, B.D. 1995. Emissions of volatile organic chemicals from municipal solid waste composting facilities. *Environ. Sci. Technol.* 29:896–902

Fielding, E.R., D.B. Archer, C. deMacario, and A.J.L. deMacario. 1988. Isolation and characterization of methanogenic bacteria from landfills. *Appl. Environ. Microbiol.* 54:835–836.

Finlay, B.J., and T. Fenchel. 1991. An anaerobic protozoon, with symbiotic methanogens, living in municipal landfill material. *FEMS Microbiol. Ecol.* 85:169–180.

Finstein, M.S., and M.L. Morris. 1975. Microbiology of municipal solid waste composting, in D. Perlman (Ed.), *Advances in Applied Microbiology*. Vol. 19. Academic Press, New York, pp. 113–151.

Fuhrman, J.A., and R.T. Noble. 1995. Viruses and protists cause similar bacterial mortality in coastal seawater. *Limnol. Oceanogr.* 40:1236–1242.

Ghosh, S., J.R. Conrad, and D.L. Klass. 1975. Anaerobic acidogenesis of wastewater sludge. *J. Water Pollut. Control Fed.* 47:30–45.

Gray, K.R., K. Sherman, and A.J. Biddlestone. 1971. A review of composting. Part I. *Process Biochem.* 6:32–36.

Harper, S.R. and F.G. Pohland. 1988. Design and management strategies for minimizing environmental impact at municipal solid waste landfill site, in Proc. 1988 Joint ASCE-CSCE National Conference on Environmental Engineering, July 13–15, 1988, Vancouver, B.C. Canada, American Society of Chemical Engineers-Canadian Society of Chemical Engineers, pp. 669–688.

He, X-T, T.J. Logan, and S.J. Traina. 1995. Physical and chemical characteristics of selected U.S. municipal solid waste compost. *J. Environ. Qual.* 24:543–552.

Huber, M.S., C.P. Gerba, M. Abbaszadegan, J.A. Robinson, and S.M. Bradford. 1994. Study of persistence of enteric viruses in landfilled disposable diapers. *Environ. Sci. Technol.* 28:1767–1772.

Iglesias-Jimenez, E., and V. Perez-Garcia. 1992. Composting of domestic refuse and sewage sludge. II. Evolution of carbon and some "humification" indexes. *Resour. Conserv. Recycl.* 8:45–60.

Johnston, R.K., W.R. Munns, Jr., L.J. Mills, F.T. Short, and H.A. Walker. 1994. Estuarine Ecological Risk Assemssment for Portsmouth Naval Shipyard, Kittery, Maine. EPA Technical Report 1627. U.S. Environmental Protection Agency, Washington, D.C.

Jones, K.L., and J.M. Grainger. 1983. The application of enzyme activity measurements to a study of factors affecting protein, starch and cellulose fermentation in domestic refuse. *Appl. Micro. Biotechnol.* 18:181–185.

Kayhanian, M. 1995. Biodegradability of the organic fraction of municipal solid waste in a high-solids anaerobic digester. *Waste Manage. Res.* 13:123–126.

Keeling, A.A., I.K. Paton, and J.A.J. Mullett. 1994. Germination and growth of plants in media containing unstable refuse-derived compost. *Soil Biol. Biochem.* 26:767–772.

Kissel, J.C., C.L. Henry, and R.B. Harrison. 1992. Potential emissions of volatile and odorous organic compounds from municipal solid waste composting facilities. *Biomass Bioenergy* 3:181–194.

Lisk, D.J. 1991. Environmental effects of landfills. *Sci. Total Environ.* 100:415–468.

Malik, M., J. Kain, C. Pettigrew, and A. Ogram.1994. Purification and molecular analysis of microbial DNA from compost. *J. Microbiol. Methods* 20:183–196.

McInerney, M.J. 1988. Anaerobic hydrolysis and fermentation of fats and proteins, in A.J.B. Zehnder (Ed.), *Biology of Anaerobic Microorganisms.* John Wiley & Sons, New York, pp. 373–400.

McKinley, V.L., and J.R. Vestal. 1985a. Effects of different temperature regimes on microbial activity and biomass in composting municipal sewage sludge. *Can. J. Microbiol.* 31:919–925.

McKinley, V.L., and J.R. Vestal. 1985b. Physical and chemical correlates of microbial activity and biomass in composting municipal sewage sludge. *Appl. Environ. Microbiol.* 50:1395–1403.

McNeil, M.B., J.M. Jones, and B.J. Little. 1991. Production of sulfide minerals by sulfate-reducing bacteria during microbiologically induced corrosion of copper. *Corrosion* 47:674–677.

Mohn, W.W., and J.M. Tiedje. 1992. Microbial reductive dehalogenation. *Microbiol. Rev.* 56:482–507.

Muyzer, G., E.C. de Waal, and A.G. Uitterlinden. 1993. Profiling of complex microbial populations by denaturing gradient gel electrophoresis analysis of polymerase chain reaction-amplified genes coding for 16S rRNA. *Appl. Environ. Microbiol.* 59:695–700.

O'Leary, P.R., P.W. Walsh, and R.K. Ham. 1988. Managing solid waste. *Sci. Am.* 259:36–42.

Pahren, H.R. 1987. Microorganisms in municipal solid waste and public health implications. *CRC Crit. Rev. Environ. Control* 17:187–228.

Palmisano, A.C., and C.A. Pettigrew. 1992. Biodegradability of plastics. *BioScience* 42:680–685.

Pavlostathis, S.G., and E. Giraldo-Gomez. 1991. Kinetics of anaerobic treatment. *Water Sci. Technol.* 24:35–59.

Peck, M.W., and D.B.Archer. 1989. Methods for the quantification of methanogenic bacteria. *Intl. Ind. Biotechnol.* 9:5–12.

Pohland, F.G. 1975. Sanitary landfill stabilization with leachate recycle and residual treatment. Georgia Insitute of Technology, EPA Grant No. R-801397.

Pohland, F.G. 1980. Leachate recycle as a landfill management option. *ASCE J. Environ. Eng.* 106:1057–1069.

Pohland, F.G. and S.R. Harper. 1987. Retrospective evaluation of the effects of selected industrial wastes on municipal solid waste stabilization in simulated landfills. EPA Project Summary EPA 600/S2-87/1044. U.S. Environmental Protection Agency, Washington, D.C.

Portney, P.R. 1993/1994. The price is right: making use of life cycle analyses. *Issues Sci. Technol.* Winter: 69–75.

Ranjithan, S., M.A. Barlaz, E.D. Brill, Jr., S-Y Fu, A. Kaneko, S.R. Nishtala, and H.R. Piechottka. Integrated solid waste management: 1. Decision support system. ASCE 2nd Congress on Computing in Civil Engineering, Atlanta, GA, June 5–7, 1995, American Society of Chemical Engineers, New York.

Reinhart, D.R. 1993. A review of recent studies on the sources of hazardous compounds emitted from solid waste landfills: a U.S. experience. *Waste Manage. Res.* 11:257–268.

Rhew, R., M.A. Barlaz. 1995. The effect of lime stabilized sludge as a cover material on anaerobic refuse decomposition. *ASCE J. Environ. Eng.* 121:499–506.

Rogers, J.E. and W.B. Whitman. 1991. Introduction, in J.E. Rogers and W.B. Whitman (Eds.), *Microbial Production and Consumption of Greenhouse Gases*. American Society for Microbiology, Washington, D.C., pp. 1–6.

Senior, E. (Ed.) 1995. *Microbiology of Landfill Sites*, 2nd ed., CRC Press, Boca Raton, FL.

Senior, E., and M.T.M. Balba. 1990. Refuse decomposition, in E. Senior (Ed.) *Microbiology of Landfill Sites*. CRC Press, Boca Raton, FL.

Sleat, R., C. Harries, I. Viney, and J.F. Rees. 1987. Activities and distribution of key microbial groups in landfill, in Proceedings of the ISWA Symposium on Process, Technology and Environmental Impact of Sanitary Landfills, International Solid Waste Association, Copenhagen.

Sobsey, M.D. 1978. Field survey of enteric viruses in solid waste landfill leachates. *Am. J. Public Health* 68:858–864.

Stenstrom, M.K., A.S. Ng, P.K. Bhunia, and S.D. Abramson. 1983. Anaerobic digestion of municipal solid waste. *J. Environ. Eng.* 109:1148–1157.

Strom, P.F. 1985a. Effect of temperature on bacterial species diversity in thermophilic solid waste composting. *Appl. Environ. Microbiol.* 50:899–905.

Strom, P.F. 1985b. Identification of thermophilic bacteria in solid waste composting. *Appl. Environ. Microbiol.* 50:906–913.

Suflita, J.M., C.P. Gerba, R.K. Ham, A.C. Palmisano, W.L. Rathje, and J.A. Robinson. 1992. The world's largest landfill: a multidisciplinary investigation. *Environ. Sci. Technol.* 26:1486–1495.

Suler, D.J., and M.S. Finstein. 1977. Effect of temperature, aeration and moisture on carbon dioxide formation in bench-scale, continuously thermophilic composting of solid waste. *Appl. Environ. Microbiol.* 33:345–350.

Tchobanoglous, G., H. Theisen, and S. Vigil. 1993. *Integrated Solid Waste Management*. McGraw-Hill, New York.

Thorneloe, S.A. and J.G. Pacey. 1994. Landfill gas utilization-database of North American projects. Solid Waste Association of North America, 17th Landfill Gas Symposium, Long Beach, CA, March 22–24, SWANA, Silver Springs, MD.

Thorneloe, S.A., S. Friedrich, M.A. Barlaz, R. Ranjithan, E.J. Kong, S. Nishtala, C. Whiles, P.B. Shepherd, and R.K. Ham. 1995. U.S. research to conduct life-cycle study to evaluate alternate strategies for solid waste management, in T.H. Christensen, R. Cossu, and R. Stegman, (Eds.), *Proc. 5th International Landfill Symposium,* Environmental Sanitary Engineering Centre, Cagliari, Italy.

Townsend, T.G., W.L. Miller, and J.F.K. Earle. 1995. Leachate-recycle infiltration ponds. *ASCE J. Environ. Eng.* 121:465–471.

USEPA. 1988. Report to Congress: Solid Waste Disposal in the United States. EPA/530-SW88-034, PB89-110381. U.S. Environmental Protection Agency, Office of Solid Waste, Washington, D.C.

USEPA.1991. Air Emissions from Municipal Solid Waste Landfills — Background Information for Proposed Standards and Guidelines. EPA-450/3-90011a. U.S. Environmental Protection Agency, Office of Air Quality Planning and Standards, Research Triangle Park, NC.

USEPA.1994a. Characterization of Municipal Solid Waste in the United States: 1994 Update. EPA/530-R-94-042. U.S. Environmental Protection Agency, Washington, D.C.

USEPA. 1994b. International Anthropogenic Methane Emissions: Estimates for 1990. EPA 230-R-93-010. U.S. Environmental Protection Agency, Office of Policy Planning and Evaluation, Washington, D.C.

USEPA. 1995. Estimate of Global Methane Emissions from Landfills and Open Dumps. EPA-600/R-95-019. U.S. Environmental Protection Agency, Air and Energy Engineering Research Laboratory, Research Triangle Park, NC.

Vallini, G., F. Cecchi, P. Pavan, A. Pera, J. Mata-Alvarez, and A. Bassetti. 1993. Recovery and disposal of the organic fraction of municipal solid waste (MSW) by means of combined anaerobic and aerobic bio-treatments. *Water Sci. Technol.* 27:121–132.

Van Meenen, P., and W. Verstraete. 1988. Anaerobic digestion of municipal solid wastes. Proceedings of the ISWA International Congress, Peruguia, Italy, 6–9 June, 1988, International Solid Waste Association, Copenhagen.

Van Meenen, P., J. Vermeulen, and W. Verstrate. 1988. Fragility of anaerobic SSF consortia, in E.R. Hall and P.N. Hobson (Eds.), *Anaerobic Digestion.* International Association of Water Pollution Research and Control, London, pp. 345–356.

Vestal, J.R. and D.C. White. 1989. Lipid analysis in microbial ecology. *BioScience* 39:535–541.

Watson-Craik, I.A., and L.R. Jones. 1995. Selected approaches for the investigation of microbial interactions in landfill sites, in E. Senior (Ed.), *Microbiology of Landfill Sites,* 2nd ed., CRC Press, Boca Raton, FL.

Westlake, K. 1989. Cellulolytic bacteria in sanitary landfill, in P.S. Lawson and Y.R. Alston (Eds.). *Proceedings of Landfill Microbiology Research and Development Workshop.* United Kingdom Atomic Energy Agency, Oxfordshire, UK.

Westlake, K., D.B. Archer, and D.R. Boone. 1995. Diversity of cellulolytic bacteria in landfill. *J. Appl. Bacteriol.* 79:73–78.

Wilkins, K. 1994. Volatile organic compounds from household waste. *Chemosphere* 29:47–53.

2

Microbiology of Solid Waste Landfills

Morton A. Barlaz

CONTENTS

2.1 Introduction 32
    2.1.1 Description of a Sanitary Landfill 33
    2.1.2 Refuse Composition 36

2.2 Biological Decomposition in a Landfill 37
    2.2.1 General Pathways of Anaerobic Decomposition 37
    2.2.2 Phases of Decomposition in Bioreactors 39
        2.2.2.1 Phase 1 — Aerobic Phase 41
        2.2.2.2 Phase 2 — Anaerobic Acid Phase 41
        2.2.2.3 Phase 3 — Accelerated Methane Production Phase 42
        2.2.2.4 Phase 4 — Decelerated Methane Production Phase 44
        2.2.2.5 Additional Observations and Summary 45
    2.2.3 The Relationship Between Laboratory and Field Observations 45

2.3 Anaerobic Bacteria Involved in Refuse Decomposition in Landfills 48
    2.3.1 Techniques for the Study of Microbiological Activity in Landfills 48
        2.3.1.1 Enumeration of Microorganisms on Refuse 49
            2.3.1.1.1 Techniques for the Formation of a Liquid
                      Inoculum from Refuse 49
            2.3.1.1.2 Microbial Enumeration 51
            2.3.1.1.3 Measurement of Anaerobic Fungi 52
            2.3.1.1.4 Habitat Simulation 52
            2.3.1.1.5 Incubation Temperature....................................52
        2.3.1.2 Direct Measures of Microbial Activity 53
            2.3.1.2.1 Enzyme Assays...53
            2.3.1.2.2 Other Measures of Microbial Activity............54
    2.3.2 Landfill Microbiology Research 55
        2.3.2.1 Pathogenic Microorganisms 58
        2.3.2.2 Microbial Populations on Individual Refuse Constituents...........58

31

2.4 Systems for the Study of Refuse Decomposition and
    Landfill Microbiology 60

2.5 Factors Limiting Decomposition in Landfills 61
    2.5.1 Regulatory Factors Influencing Methane Production 63

2.6 Relationship Between Biological Decomposition and Leachate
    and Gas Characteristics..63

2.7 Future Research Directions..65

Acknowledgments 66

References 66

2.1 INTRODUCTION

A landfill is an ultimate disposal alternative for municipal solid waste (MSW). Development of integrated solid waste management programs, which include recycling, yard waste composting, and/or combustion, has led to a decrease in the fraction of MSW directed to landfills (USEPA, 1994a). However, there is a limit to the types of waste that can be recycled, and combustion is not the waste management alternative of choice for many communities. Thus, landfills will be a significant part of MSW management for the foreseeable future. In 1993, approximately 62% of the MSW generated in the U.S. was disposed of by burial in a sanitary landfill, a decrease of 23% relative to 1980 (USEPA, 1994a).

A complex series of microbiological and chemical reactions begins with the burial of refuse in a landfill, and the production of methane and carbon dioxide from landfills is well documented (Ham et al., 1979; Emcon Associates, 1980). Methane production from landfills has several important ramifications. Methane represents a recoverable source of biomass-derived energy, and in 1992 methane was recovered in commercial quantities at 119 landfills in the U.S. and Canada (Thorneloe and Pacey, 1994). Typically, this methane is converted to electricity, although in some cases it is piped directly into industrial boilers or cleaned up to natural gas pipeline standards. In addition to the benefits of methane recovery for energy, there are other environmental implications of methane production. First, both methane and carbon dioxide are greenhouse gases that contribute to global climate change (USEPA, 1990). While these gases are produced in approximately equal proportions in landfills, methane is estimated to be at least 20 times more damaging than carbon dioxide on a volume basis (USEPA, 1994b). Second, leachate strength is reduced with the onset of methane production. This lessens both the risk of groundwater contamination and the cost of leachate treatment. Finally, methane production is coupled to refuse decomposition; the resulting decrease in refuse volume in turn increases the volume of landfill space available.

The objectives of this chapter are to (1) describe the microbiological and chemical reactions involved in refuse decomposition in landfills and the relationship of these reactions to other aspects of landfills such as gas production and leachate composition and (2) review research on the anaerobic microorganisms involved in refuse decomposition. As methane production is a microbiologically mediated process, an understanding of the microbiology of landfills will result in better predictive models of methane production and the potential to manipulate the system for increased energy production. Much of the initial research on landfill microbiology has focused on counts of fecal bacteria and pathogens (Cook et al., 1967; Donnely and Scarpino, 1984; Kinman et al., 1986). Such work is useful for evaluation of potential health problems associated with refuse and landfill leachate but does not elucidate the microbial processes responsible for landfill methanogenesis. In the early 1980s, researchers in the United Kingdom began to focus on the anaerobic processes associated with methane production in landfills (Jones and Grainger, 1983; Jones et al., 1983), and there is now a body of literature on anaerobic processes involved in refuse decomposition. Landfills represent a tremendous challenge for microbiological research as their size, contents, and heterogeneity require the use of novel methods for sample recovery and analysis. Nevertheless, as summarized in this chapter, techniques have been developed, substantial progress has been made in the study of landfill microbiology, and good descriptions of the microbiological and chemical reactions that occur in landfills are available.

2.1.1 Description of a Sanitary Landfill

While landfills historically have been the dominant alternative for the ultimate disposal of MSW, there has been substantial evolution in their design and operation. In the past, a landfill often represented little more than an open hole or marsh where refuse was dumped. The refuse often was not covered properly, sometimes it was burned for volume reduction, and there was little effort to control stormwater runoff and downward migration of water that had come into contact with the refuse (leachate). As late as 1986, an estimated 6000 municipal landfills were operating, in addition to thousands of closed facilities (USEPA, 1988). Most of these sites were constructed without an engineered liner or leachate collection system, and nearly 50% were within 1 mile of a drinking-water well. With the implementation of Subtitle D of the Resource Conservation and Recovery Act (RCRA) and regulatory programs in many states, permits are now required for the construction and operation of a landfill. As such, the permitted landfill of today is a highly engineered facility designed to contain the refuse and separate it from the environment, to capture leachate, and to control methane migration. Moreover, modern landfills tend to be larger, with wastes transported from an entire region as opposed to a single city or county.

The major components of a landfill are briefly summarized here and illustrated in Figure 1. Detailed information on landfill design is presented in books by Pfeffer (1992), Tchobanoglous et al. (1993), and Daniel (1993). Prior to placement of waste in the ground, a site is typically excavated to increase the available disposal volume per acre. The lowermost component of a landfill is the liner system which includes layers to (1) minimize the migration of leachate to the groundwater, and (2) collect leachate for treatment. A common system used to minimize leachate migration begins with a layer of low permeability soil, typically a 0.67- to 1-m thick clay layer with a hydraulic conductivity of no more than 10^{-7} cm/sec. A flexible membrane liner (FML) is often placed above the clay layer. The FML is typically 1.5-mm thick polyethylene with an equivalent hydraulic conductivity (based on vapor diffusion) of about 10^{-11} cm/sec. Together, the soil and FML are referred to as a *composite liner,* whereas the clay layer alone would be referred to as a *clay liner.* A drainage layer designed to promote the collection of leachate is placed above the composite or clay liner. This layer consists of a sand with a permeability of at least 10^{-2} cm/sec. Slotted pipe is placed in the sand layer at intervals of 30 to 60 m to collect leachate and route it to a treatment system. A system that includes a composite liner and a leachate collection system meets current federal design regulations for the bottom of a landfill. Some states have stricter design standards which may include a second liner, either clay or composite, and a second leachate collection system which is designed to detect leachate which migrates through the primary liner. A review of alternate liner systems has been presented by Fluet et al. (1992).

A barrier is installed above the leachate collection system to protect it from the equipment used to place refuse in a landfill. Examples of such barriers include additional soil, baled waste, or tire chips. Refuse may then be placed above the protective layer. Trucks delivering the waste contain residential and/or commercial waste; thus, there is much heterogeneity in the waste stream as buried. The waste is compacted and covered daily to minimize the attraction of rodents, wildlife, and disease-carrying insects; blowing of the refuse away from the landfill; and contamination of stormwater runoff. Traditionally, a 15-cm soil layer was used for daily cover, although alternatives are becoming more common. Alternatives include (1) synthetic materials that are rolled over the waste at the end of the working day and are removed prior to burial of additional refuse, and (2) foams which cover the waste but collapse when additional waste is placed over them. The refuse received and covered in a day is referred to as a *daily cell.*

Landfills typically are designed to receive refuse for at least 10 and up to 50 or more years; they are built in sections with about a 5-year capacity. Refuse is buried in landfills in accordance with detailed fill plans. Once refuse has reached the final design grade of the landfill, a final cover is applied. At a minimum, this should include a layer of low permeability soil designed to minimize stormwater infiltration, overlaid by a layer that will support vegetative

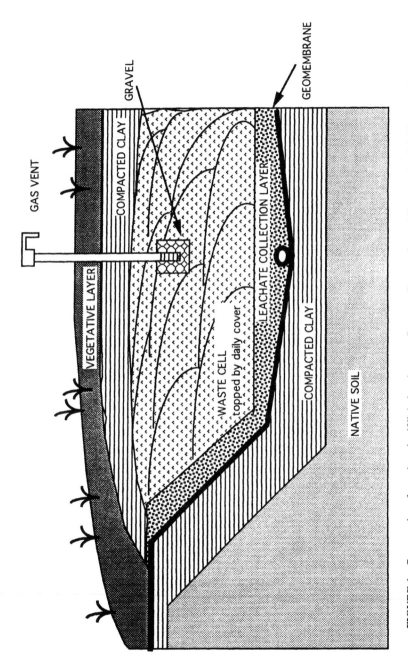

FIGURE 1 Cross section of a sanitary landfill designed to meet Resource Conservation and Recovery Act Subtitle D specifications.

growth. Vegetation serves to minimize erosion of the final cover and promote evapotranspiration. The use of final covers that more effectively retard infiltration is becoming more common. Such a cover could include a layer of clay over the refuse, followed by a drainage layer, and then a layer to support vegetative growth. The drainage layer will capture water that has percolated through the vegetative layer and will minimize its further percolation toward refuse.

Two additional aspects of landfill design related to refuse biodegradation are gas production and leachate quality. Methane and carbon dioxide are the terminal products of anaerobic decomposition in landfills. These gases typically are vented through wells placed in the refuse in order to minimize their migration off-site. The gas may be vented, flared to reduce the release of methane, or recovered for use as energy. Unsaturated zone monitoring wells around the site are required to verify that methane is not migrating off-site at potentially dangerous concentrations. Groundwater monitoring wells must also be installed around the site to verify that leachate is not bypassing the leachate collection system and liner in significant quantities.

2.1.2 Refuse Composition

Traditionally, MSW has been classified according to visual categories such as glass, paper, and metals, as described in Table 1 in Chapter 1. While such a characterization is useful for recycling studies, the organic composition of MSW is more useful for decomposition studies. Data on the organic composition of refuse and certain of its subcomponents are presented in Table 2. These data indicate that cellulose and hemicellulose are the principal biodegradable components of MSW. The other major organic component of MSW, lignin, is recalcitrant under anaerobic conditions (Young and Frazer, 1987). In addition to its recalcitrance, lignin can interfere with the decomposition of cellulose and hemicellulose by physically impeding microbial access to these degradable carbohydrates. Though recalcitrant, the chemical form of lignin may change during decomposition, resulting in apparent changes in the lignin concentration, as measured by the 72% sulfuric acid digestion procedure (Effland, 1977; Iiyama et al., 1994). Other biodegradable organic compounds present in smaller concentrations are protein and soluble sugars. Protein concentrations typically are obtained by multiplication of the organic nitrogen content by 6.25. In addition to protein, other forms of organic nitrogen are likely to be in refuse which may be less degradable, such as structural materials and nitrogen-containing humic compounds. Thus, the estimated protein concentrations are likely to represent the upper limit on the actual percentage of protein in refuse. Where measured, pectin and soluble sugars were present in very small concentrations (Barlaz et al., 1990). The data in Table 1 indicate that the methane potential of a landfill is controlled by the input of refuse components rich in cellulose and hemicellulose.

TABLE 1   Cellulose, Hemicellulose, and Lignin Concentration of Individual
Waste Components (Percent of Dry Weight)

Component	Cellulose	Hemicellulose	Lignin	Volatile Solids
Grass	26.5	10.2	28.4	85.0
Leaves	15.3	10.5	43.8	90.2
Branches	35.4	18.4	32.6	96.7
Food waste[a]	50.8	6.7	9.9	92.0
Office paper	87.4	8.4	2.3	98.6
Coated paper	42.3	8.4	15.0	74.3
Newsprint	48.5	9.0	23.9	98.5
Corrugated boxes	57.3	9.9	20.8	98.2
Mixed refuse	51.2[b]	11.9[b]	15.2[b]	78.6[b]
	28.2[c]	9.0[c]	23.1[c]	75.2[c]
	48.2[d]	10.6[d]	14.5[d]	

[a] Data are the average of two samples, each of which was a composite of food waste collected in five kitchens over a 7-day period. The protein content (6.25 × total Kjeldahl nitrogen, TKN) of one of the two samples was 18.8%.

[b] Collected in Madison, WI, in 1987. The protein content (6.25 × TKN) was 4.2%.

[c] Collected in Raleigh, NC, in 1992.

[d] Collected in Raleigh, NC, in 1994.

2.2 BIOLOGICAL DECOMPOSITION IN A LANDFILL

The decomposition of MSW to methane in sanitary landfills is a microbially mediated process that requires the coordinated activity of several trophic groups of bacteria. As discussed above, the principal biodegradable substrates are cellulose and hemicellulose. Similar substrates fuel the production of methane in other ecosystems including the rumen, marshes, rice paddies, and sludge digesters (Wolfe, 1979). The general pathway for anaerobic decomposition, as it has been documented to occur in other anaerobic ecosystems, is reviewed in the first part of this section, followed by presentation of data specifically applicable to landfills. This section concludes with a discussion of the relationship between laboratory measurements and full-scale landfills.

2.2.1 General Pathways of Anaerobic Decomposition

Several trophic groups of anaerobic bacteria, each performing a narrow class of reactions, are required for the production of methane from biological polymers such as cellulose, hemicellulose, and protein (Zehnder, 1978) (Figure 2). The first reaction is the hydrolysis of polymers, including carbohydrates, fats, and proteins. The initial products of polymer hydrolysis are soluble sugars, amino acids, long-chain carboxylic acids, and glycerol. Fermentative microorganisms then ferment these hydrolysis products to short-chain carboxylic acids, carbon dioxide, and hydrogen. Acetate, a direct precursor of methane,

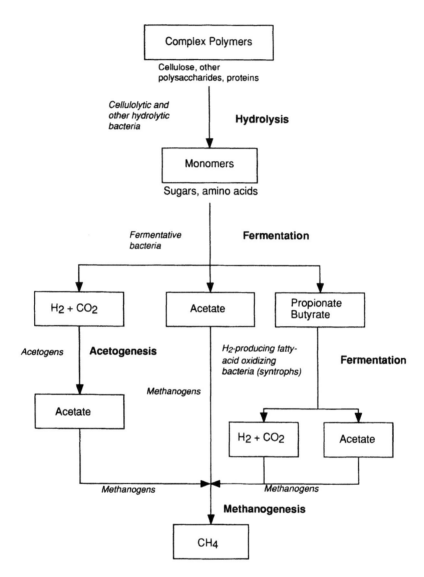

FIGURE 2 Overall process of anaerobic decomposition showing the conversion of complex
organic materials to methane and carbon dioxide by consortia of anaerobic bacteria.
(Redrawn from Brock, T.D., M.T. Madigan, J.M. Martinko, and J. Parker. 1994.
Biology of Microorganisms, 7th ed., Prentice Hall, Englewood Cliffs, NJ, p. 650.
With permission.)

and alcohols are also formed. The next reaction in the conversion of biological
polymers to methane is carried out by bacteria referred to as *obligate proton-
reducing* (or H_2-producing) *acetogens*. They oxidize fermentation products,
including propionate and butyrate, to acetate, carbon dioxide, and hydrogen.
Oxidation of propionate and butyrate is only thermodynamically favorable at

very low hydrogen concentrations (Zehnder, 1978). Thus, the obligate proton-reducing acetogenic bacteria only function in syntrophic association with a hydrogen scavenger such as a methanogen or a sulfate reducer. Typically, sulfate concentrations in landfills are minimal and methane is the major electron sink, although exceptions exist. The significance of another acetogenic reaction, the production of acetate from H_2 and CO_2, has not been established in the MSW landfill ecosystem. This microbiological activity probably competes weakly with the hydrogenophilic methanogens for hydrogen.

The terminal step in the conversion of complex polymers to methane is carried out by methanogenic bacteria. Methanogens convert either acetate or H_2 plus CO_2 to methane. In sludge digesters it is estimated that 70% of the methane produced originates from acetate (Zeikus, 1980). This value has not been investigated in the landfill ecosystem. The production of methane from acetate yields only 31 kJ per mole CH_4 produced. This is barely enough energy for the generation of adenosine triphosphate (ATP), which requires 30.6 kJ/mole. Thus, the growth of methanogens on acetate is relatively slow. The conversion of H_2 plus CO_2 to CH_4 yields 135.6 kJ per mole CH_4 produced. Thus, the latter reaction is energetically more favorable. The methanogens are most active in the pH range 6.8 to 7.4 (Zehnder, 1978).

The importance of the methanogens in anaerobic digestion has been summarized by Zeikus (1980). As a group, the methanogens (1) control the pH of their ecosystem by the consumption of acetate, and (2) regulate the flow of electrons by the consumption of hydrogen, creating thermodynamically favorable conditions for the catabolism of alcohols and acids. Should the activity of the fermentative organisms exceed that of the acetogens and methanogens, there will be an imbalance in the ecosystem. Carboxylic acids and hydrogen will accumulate, and the pH of the system will fall, thus inhibiting methanogenesis.

2.2.2 Phases of Decomposition in Bioreactors

When refuse is placed in a landfill, biological decomposition resulting in methane formation as described in the previous section does not occur immediately. A period ranging from months to years is necessary for the proper growth conditions and the required microbiological system to become established. Thus, most research on refuse decomposition has been conducted using laboratory simulations in which the rate of decomposition was accelerated. A characterization of refuse decomposition that describes chemical and microbiological characteristics was developed by Barlaz et al. (1989a) using data from a set of laboratory-scale reactors. Refuse decomposition is described in an aerobic phase, an anaerobic acid phase, an accelerated methane production phase, and a decelerated methane production phase. Each phase is described below and summarized in Figure 3. Following presentation of this laboratory-derived decomposition cycle, its relationship to decomposition characteristics of a full-scale landfill is discussed.

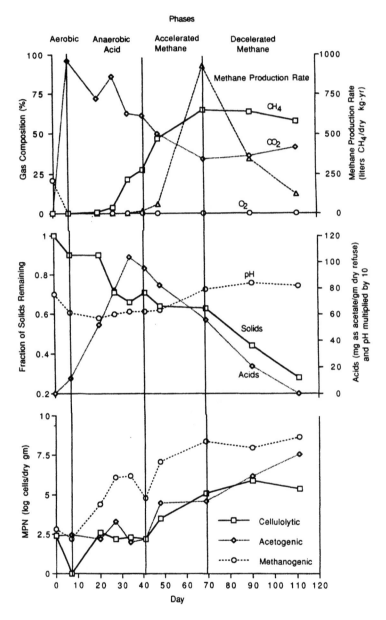

FIGURE 3 The four phases of refuse decomposition in laboratory-scale reactors simulating a
sanitary landfill. Gas volume data were corrected to dry gas at standard temperature
and pressure. The acids are expressed as acetic acid equivalents. Solids remaining
equal the ratio of the cellulose plus hemicellulose removed from a reactor divided
by the weight of cellulose plus hemicellulose added to the reactor initially. Methanogen most probable number (MPN) data are the log of the average of the acetate-
and H_2/CO_2-utilizing populations. (Redrawn from Barlaz et al. 1989a. *Appl. Environ.
Microbiol.*, 55:60. With permission.)

2.2.2.1 Phase 1 — Aerobic Phase

Oxygen is entrained in the void space when refuse is buried. This oxygen, plus oxygen dissolved in the refuse-associated moisture, supports aerobic decomposition. In the aerobic phase, oxygen is consumed, with soluble sugars serving as the carbon source for microbial activity. Gas composition will be nearly 100% CO_2 during the aerobic phase. Though not considered aerobic metabolism, nitrate is also rapidly consumed soon after refuse burial and is considered as part of the aerobic phase. As illustrated in Figure 4, all of the trophic groups required for refuse methanogenesis (cellulolytics, acetogens, and methanogens) have been measured in fresh refuse at populations of 100 to 1000 cells/dry g. Population enumerations on decomposed samples are discussed below. The pH of fresh refuse was about neutral; however, as described in phase 2, it declines rapidly as oxygen is depleted and fermentation begins.

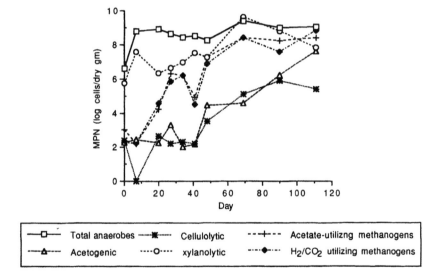

FIGURE 4 Changes in microbial populations during refuse decomposition in laboratory-scale reactors simulating a sanitary landfill. (Redrawn from Barlaz et al. 1989a. *Appl. Environ. Microbiol.*, 55:62. With permission.)

2.2.2.2 Phase 2 — Anaerobic Acid Phase

The anaerobic acid phase begins following the depletion of oxygen and nitrate from the refuse ecosystem. Rapid oxygen depletion was measured in the laboratory and can be expected in the field because once a mass of refuse is covered by other refuse, replenishment of oxygen becomes insignificant. The acid phase is characterized by the rapid accumulation of carboxylic acids and a decrease in refuse pH from around neutral to below 6.0. This pH decrease is due to the accumulation of acidic intermediates of sugar fermentation which

results from the low acid-consuming activities of the acetogenic and methanogenic populations. The total anaerobic population exceeds the methanogenic
and acetogenic populations by a factor of 10^6 at the beginning of the anaerobic
acid phase. In addition to carboxylic acids, a wide variety of organic intermediates is likely to be produced. High CO_2 concentrations result from fermentative activity. Although H_2 concentrations are not presented in Figure 3, they
were detected prior to the onset of methane production (Barlaz et al., 1989a).
Both CO_2 and H_2 are products of sugar fermentation.

Carboxylic acid concentrations reach a maximum and pH reaches a minimum during the anaerobic acid phase. Despite the acidic conditions which
dominate the anaerobic acid phase, microbial population development does
occur. The total anaerobic and hemicellulolytic (as represented by xylan-
degrading microbes) populations increase by factors of 100 and 10, respectively, between the aerobic phase and early in the anaerobic acid phase (Figure
4). Thereafter, these populations remain nearly constant. Both the acetate-
utilizing (acetoclastic) and H_2/CO_2-utilizing (hydrogenophilic) methanogen
populations increase by four orders of magnitude during the anaerobic acid
phase, although the pH ranges from 5.7 to 6.2, which is well below the optimal
pH for methanogenesis. There is not a sustained increase in the acetogen or
cellulolytic populations during the anaerobic acid phase.

With progression through the acid phase, the CO_2 concentration decreases
as the CH_4 concentration increases. The CH_4 concentration increase prior to
an increase in the methane production rate is a reflection of the gas volume
detection limits in the laboratory-scale system (see Figure 3). Cellulose and
hemicellulose hydrolysis is not consistent in the acid phase and there is
relatively little solids hydrolysis. External neutralization in the laboratory
system improved the conditions for acetogenic and methanogenic activity and
propelled the ecosystem out of the acid phase and into the accelerated methane
production phase described below.

2.2.2.3 Phase 3 — Accelerated Methane Production Phase

In the accelerated methane production phase, there is a rapid increase in the
rate of methane production to its maximum value. Methane concentrations of
50 to 70% are typical of this phase, with the balance of the gas being CO_2.
Carboxylic acid concentrations decrease sharply in the third phase of refuse
decomposition. They are consumed faster than they are produced as the methane production rate increases; subsequently, the pH of the refuse ecosystem
increases. There is little solids hydrolysis during this phase of decomposition
(see Figure 3).

There is little change in the total anaerobic and hemicellulolytic populations
during phase 3. Methanogen populations increase by one to two orders of
magnitude. Most notably, the cellulolytic and acetogenic populations increase
relative to their numbers in fresh refuse.

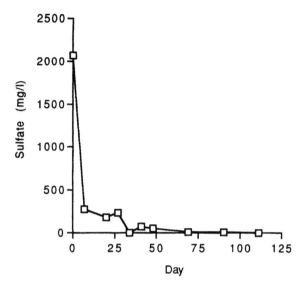

FIGURE 5 Sulfate concentrations during refuse decomposition in laboratory-scale reactors simulating a sanitary landfill.

The sulfate concentration remains high in phases 1 and 2 and decreases thereafter (Figure 5). The high concentrations may be due to the recycling of acidic leachate. Data suggest that the refuse ecosystem has a high capacity to reduce sulfate. An observed decline in sulfate concentration after day 69 does not necessarily mean that the rate of sulfate reduction has decreased; sulfate may have been solubilized by acidic leachate and then immediately reduced to hydrogen sulfide. Sulfate-reducing bacteria have the capacity to outcompete methanogenic bacteria for hydrogen (Robinson and Tiedje, 1984). Thus, in a hydrogen-limited ecosystem, sulfate will inhibit methane production, although some methane production will occur (Robinson and Tiedje, 1984). The high production of carboxylic acids suggests that hydrogen may not be limiting in the refuse ecosystem. Note that the methane production rate in Figure 3 was increasing on day 41 in the presence of 0.5 mm SO_4 per liter (0.5 mM). Thus, it is unlikely that sulfate was inhibitory in this system. Nevertheless, higher concentrations of sulfate could divert electron flow from methane production to sulfate reduction.

In the accelerated methane production phase, the methanogenic bacteria appear to reach a critical level. They begin to pull the refuse fermentation by consumption of acetate and hydrogen. Hydrogen consumption improves conditions for the conversion of butyrate and propionate to acetate. Acetate consumption allows the pH of the ecosystem to increase. As the pH increases, the acetogen population increases, which allows for additional carboxylic acid consumption and a further increase in the methane production rate. At the end of this phase, the methanogenic bacteria comprise 10% of the total anaerobic population.

2.2.2.4 Phase 4 — Decelerated Methane Production Phase

The final phase of refuse decomposition is described as the decelerated methane production phase. It is characterized by a decrease in the methane production rate while the CH_4 and CO_2 concentrations remain constant at about 60 and 40%, respectively. The decrease in the methane production rate correlates with a decrease in carboxylic acid concentrations which drop to below 100 mg/l. With the depletion of carboxylic acids, there is a further increase in the pH of the ecosystem. The rate of cellulose and hemicellulose decomposition in the decelerated methane production phase is higher than that exhibited in any other phase of refuse decomposition. Solids hydrolysis controls the rate of methane production as there is no longer an accumulation of carboxylic acids to serve as soluble substrate. The total anaerobic, hemicellulolytic, and methanogenic populations did not exhibit significant changes in phase 4; however, the acetogen population continued to increase.

In earlier phases of refuse decomposition, the hydrolysis of cellulose and hemicellulose resulted in accumulations of carboxylic acids. In the fourth phase of decomposition, where the rate of polymer hydrolysis exceeds that exhibited earlier, no accumulations are observed. The difference in carboxylic acid accumulation between phases 2 and 4 may be explained by the relative strengths of the hydrolytic/fermentative, methanogenic, and acetogenic populations. In phase 2, the fermentative population produces acids faster that they can be consumed. Later, the acetogenic and methanogenic populations are high enough to consume the acids as they are produced.

There has been speculation that high concentrations of carboxylic acids inhibit methane production; however, acetate and butyrate concentrations of 9753 and 6956 mg/l, respectively, did not inhibit methane production in the reactors described by Figure 3 (Barlaz et al., 1989c). In the second and third phases of refuse decomposition, it is the utilization of carboxylic acids that limits the onset and rate of methane production. After consumption of the initial accumulation of carboxylic acids, it is polymer hydrolysis which limits the rate of methane production.

The four-phase characterization of refuse decomposition presented here represents what would happen under ideal circumstances for a given volume of newly placed refuse. The data presented in Figure 3 have time as the abscissa. The times shown on the abscissa should not be applied to other conditions as these times were influenced by (1) the daily recycling and neutralization of leachate, (2) the incubation temperature (40°C), and (3) the use of shredded refuse. Nonetheless, no external additions of bacteria were made to the refuse. It is believed that the major differences in refuse decomposition under different conditions will be the length of time required for the different phases of refuse decomposition to evolve, the methane production rate, the extent of biodegradation, and the methane yield. The relationship between the laboratory data presented here and full-scale landfills is discussed in Section 2.2.3.

2.2.2.5 Additional Observations and Summary

Several observations concerning refuse decomposition and the landfill ecosystem can be made based on an understanding of the groups of microorganisms involved. The methanogens are most active in the pH range 6.8 to 7.4 (Zehnder, 1978). Should the activity of the fermentative organisms exceed that of the acetogens and methanogens, there will be an imbalance in the ecosystem. Carboxylic acids and hydrogen will accumulate and the pH of the system will fall, thus inhibiting methanogenesis. This is what happens in the anaerobic acid phase. The potential for a rapid accumulation of acids is particularly acute in the refuse ecosystem. Food and yard wastes make up about 23% of MSW (see Chapter 1, Table 1) and these materials contain soluble sugars. These sugars will be converted to carbon dioxide in the presence of oxygen or nitrate or carboxylic acids in the absence of oxygen and nitrate. In one fresh refuse sample, the measured concentrations of soluble sugars and nitrates were reported to be 3.46 and 0.015 mg/dry g, respectively. Assuming a porosity of 25%, Barlaz et al. (1989a) calculated that the amounts of oxygen and nitrate present in freshly buried refuse were sufficient to support the oxidation to carbon dioxide of only 8% of fresh refuse soluble sugars. This means that 92% of the sugars are fermented to carboxylic acids, which will accumulate if not further metabolized, reducing the pH of the refuse ecosystem and inhibiting methanogenesis.

All soluble organic chemicals exert a chemical oxygen demand (COD). Prior to the decelerated methane production phase, carboxylic acids accounted for 60 to 90% of the leachate COD (Barlaz et al., 1989a). Thus, high organic acid concentrations and high COD may be considered equivalent. The high organic acid concentrations plus dissolved carbon dioxide cause acidic leachate that increases dissolution of inorganic constituents. Moreover, naturally occurring organic compounds can act as chelating agents and dissolve metals that would otherwise be insoluble.

2.2.3 The Relationship Between Laboratory and Field Observations

The four phases of refuse decomposition described in Section 2.2.2 are based on a laboratory-scale system in which the leachate was neutralized and recycled and the temperature was controlled to accelerate the onset of methane production. While this enhancement was useful for elucidation of a theoretical description of refuse decomposition, it is important to understand the relationship between the laboratory observation and what occurs in full-scale landfills where leachate recycling and neutralization typically is not practiced.

Perhaps the most important difference between laboratory and field measurements is that, when a laboratory system is used, the refuse is well defined and its solid, liquid, and gaseous characteristics can all be related to a measured state of decomposition and methane production. In contrast, gas and leachate samples recovered from a full-scale landfill do not represent refuse in a specific state of decomposition, but rather they represent composite samples of refuse which could be in a number of different phases of decomposition. Solid samples typically are not collected from full-scale landfills due to the expense of sample collection and the difficulties associated with refuse heterogeneity.

The aerobic phase in a full-scale landfill is likely to last for only a few days. During the aerobic phase, refuse usually is not saturated with water. Thus, any leachate produced from freshly buried refuse likely will have preferentially flowed through a channel in the refuse. Leachate generated from freshly buried refuse has been analyzed from test landfills and laboratory cells. This leachate has sometimes been observed to be very highly concentrated, with COD values of 10^4 to 10^5 mg/l. Such numbers primarily reflect squeezings from refuse compaction. Leachate released from fresh refuse likely will pass through refuse in a more advanced state of decomposition before collection at the bottom of a landfill. Thus, leachate characteristic of the aerobic phase may not be observed. Instead, liquid percolating through fresh and then older refuse will reflect the characteristics of the older refuse.

The anaerobic acid phase explains the long lag time between refuse burial and the onset of methane production in sanitary landfills in that low pH inhibits refuse methanogenesis. Leachate representative of the anaerobic acid phase of decomposition is acidic and has high COD and carboxylic acid concentrations. Thus, the COD or total organic carbon (TOC) of landfill leachate is typically highest before the onset of steady-state methanogenesis (Ham and Bookter, 1982). If leachate representative of the anaerobic acid phase percolates through methanogenic refuse prior to collection, then the pH is likely to increase and the COD is likely to decrease as acids are consumed by the methane-producing refuse. Thus, the failure to observe acidic leachate does not necessarily indicate that a portion of the refuse in a landfill is not in the acid phase.

As refuse enters the accelerated methane production phase, carboxylic acids are converted to CH_4 and CO_2; thus, the accelerated methane production phase is characterized by decreasing carboxylic acid concentrations. However, this decrease is relative and no absolute carboxylic acid concentrations can be provided to distinguish between refuse in the anaerobic acid and accelerated methane production phases of decomposition. The sharp increase and decrease illustrated in Figure 3 will be considerably dampened in the field due to the influence of refuse in varying states of decomposition. In addition to the state of decomposition, leachate strength or concentration is influenced by external events such as dilution by stormwater infiltration. Finally, if leachate containing carboxylic acids passes through refuse in the decelerated phase of decomposition, where solids hydrolysis limits the rate of methane production, these acids are likely to be consumed prior to collection at the bottom of a landfill.

Carboxylic acids are essentially depleted in the decelerated methane production phase, with concentrations reflective of steady-state pool levels. The steady-state concentration has not been established for decomposing refuse. Though the acids are depleted, there is still some COD exerted by the leachate.

While relatively low, it is likely to consist of less degradable compounds, including humic materials.

In addition to the organic components of leachate discussed above, leachate will also include inorganic constituents, the solubility of which are influenced by its pH and carboxylic acid concentration. Refuse pH decreases during the anaerobic acid phase. The acidic pH results in a chemically aggressive leachate which increases the dissolution of inorganic constituents including salts and heavy metals. In addition to solubility effects, carboxylic acids also act as chelating agents (Francis and Dodge, 1986) which increase metals concentrations. As refuse decomposition proceeds to the decelerated methane production phase, humic materials are produced. These materials also behave as chelating agents. In an evaluation of the co-disposal of metal sludges with MSW, Pohland and Gould (1986) reported an increase in metal concentrations in leachate as the refuse became well decomposed.

Metal concentrations will also be influenced by the presence of sulfate. Sulfate reduction to sulfide will decrease metals solubility, as most metal sulfides are extremely insoluble. Conceivably, sulfide produced in one section of a landfill could be transported to another section. The pK_a of hydrogen sulfide is 7.2. Thus, at pH 7.2, 50% of the total dissolved sulfide will be in the H_2S form from which it may volatilize. Volatilization would be enhanced by active gas production which would strip H_2S from the aqueous phase, facilitating its transport in, and removal from, a landfill. In summary, heavy metal concentrations in leachate are a result of a number of chemical processes. The presence of high sulfate concentrations in refuse has the potential to influence metals concentrations in leachate by production of sulfide, a sink for metals.

Methane and carbon dioxide concentrations and methane production characteristics of each of the four phases of decomposition are summarized in Figure 3 and were discussed in Section 2.2.2. As with leachate, gas collected at a landfill is representative of a composite sample over a wide area of buried refuse. The aerobic phase of decomposition, characterized by high CO_2 concentrations, is short and is restricted to fresh refuse that is not likely to comprise a significant fraction of the refuse in a landfill. In addition, gas venting or extraction wells typically are not installed until a section of a landfill is capped, by which time refuse decomposition will have proceeded beyond the aerobic phase. Only minor amounts of gas are produced during the anaerobic acid phase. Thus, gas composition from a landfill is not likely to reflect refuse in the anaerobic acid phase. Methane concentrations in the accelerated and decelerated phases of decomposition are generally about 60% CH_4 and 40% CO_2, and the composition of gas produced in landfills is typically in this range. The presence of oxygen and/or nitrogen probably reflects the intrusion of atmospheric gases into the landfill as a result of gas extraction at a rate which exceeds the rate of gas production. Under field conditions, a constant methane production rate is typically observed between phases 3 and 4. This probably reflects the fact that the measured methane production rate in a full-scale landfill is a composite rate developed from numerous areas with unique methane production characteristics.

The major sources of methane production from landfills are cellulose and hemicellulose. As these components are converted to gas, the volume of solid material is reduced, resulting in settlement. In the study summarized in Figure 3, 72% of the initial cellulose plus hemicellulose was decomposed over 111 days. Of course, this rate of solids decomposition exceeds that which can be expected in the field. While there are few data on the rate of settlement, it does occur in full-scale landfills and must be addressed during the design phase so that leachate and gas collection systems do not break. In addition, settlement of the final cover after placement on a landfill can result in cracks in the cover which allow for increased stormwater infiltration and gas release.

In summary, laboratory measurements have provided a good description of the microbial conversion of refuse to methane and carbon dioxide. In the next section, we begin to look in more detail at the manner in which the microbial reactions and the microorganisms pertaining to refuse decomposition can be studied.

2.3 ANAEROBIC BACTERIA INVOLVED IN REFUSE DECOMPOSITION IN LANDFILLS

The previous sections of this chapter have focused on the biological reactions that occur in landfills and their influence on parameters such as pH, COD, and methane production. In this section, work on the measurement of microbial activity is reviewed. This section begins with a discussion of applicable techniques, including enumeration of microbial populations and measurement of enzyme activities. Recent literature on the presence and activity of the anaerobic microorganisms which participate in refuse decomposition is then reviewed.

2.3.1 Techniques for the Study of Microbiological Activity in Landfills

A discussion of techniques applicable to landfill microbiology must begin with an acknowledgment that landfills are extremely heterogeneous due to the nature of the waste buried therein. It probably is not possible to obtain "truly representative" samples from landfills. Nonetheless, by collection of numerous samples from different parts of a landfill, one can obtain some indication of the status of the landfill. Given the heterogeneity issue, some researchers have studied landfill microbiology in laboratory-scale simulations, while others have focused on field sampling. Alternate systems for the study of landfill microbiology and refuse decomposition are discussed in Section 2.4.

Measurements that can be used to quantify microbial activity in refuse include enumerations by most probable number (MPN) tests, viable counts using petri dishes or roll tubes, total counts using acridine orange staining techniques, and enzyme assays. Each of these techniques will be discussed in this section.

All culture techniques have a common limitation which should be recognized prior to their application. When culture techniques are used to enumerate populations in environmental samples, it is assumed that the desired populations will grow under the laboratory conditions provided. However, this assumption is not completely accurate, as demonstrated by data from the Fresh Kills landfill where the ratio of the total fermentative population measured in roll tubes to the acridine orange direct count (AODC) was between 2.2×10^{-5} and 0.016 (Palmisano et al., 1993b). This result is typical of other ecosystems and serves to emphasize the limitations of laboratory culture techniques. The development of alternate techniques to enumerate more accurately the viable bacteria in different ecosystems is an area of active research (Winding et al., 1994).

2.3.1.1 Enumeration of Microorganisms on Refuse

Techniques that have been used for the enumeration of bacteria include MPNs (Barlaz et al., 1989a; Suflita et al., 1992), AODCs (Palmisano et al., 1993b), and roll tubes (Westlake and Archer, 1990; Palmisano et al., 1993b). Regardless of the enumeration technique, the first step must be the formation of a liquid inoculum from a refuse sample. Inoculum formation techniques are discussed first, followed by a discussion of specific enumeration techniques.

2.3.1.1.1 Techniques for the Formation of a Liquid Inoculum from Refuse. The only published study in which alternate techniques for inoculum formation were evaluated is that by Barlaz et al. (1989b). The basic technique was to blend the refuse in anaerobic phosphate buffer in a blender with a 4-liter capacity. The blended material was then squeezed by hand and the squeezings were used as the inoculum. Treatments evaluated to increase the efficiency of cell extraction included prechilling the refuse at 4°C, multiple blendings and hand squeezings, and the use of blended refuse prior to hand squeezing. These treatments were based on a survey of the rumen microbiology literature where there has been more extensive work on the enumeration of bacteria attached to cellulosic substrates. The additional treatments did not increase the MPNs of cellulolytic bacteria above the population measured by blending followed by hand squeezing.

In further work, the inoculum formation technique was validated by addition of a spike of rumen fluid to refuse followed by recovery of the spiked cellulolytic bacteria using the inoculum formation technique and MPN enumerations. Results showed no evidence that the refuse was exerting a toxic effect on the ruminal cellulolytic bacteria or that these bacteria were irreversibly attaching to refuse.

The use of a blender was appropriate in the work of Barlaz et al. (1989a) because all work was conducted with shredded refuse. Using a blender it was possible to blend approximately 100 g (dry weight) of refuse. Palmisano et al.

(1993b) worked with 10-g samples of refuse excavated from landfills. Samples were placed in a stomacher with prereduced anaerobic buffer and mixed for approximately 1 minute. Westlake and Archer (1990) added 1-g samples to Hungate tubes containing anaerobic diluent and agitated the tubes to form an inoculum. The use of small samples suggests that there is some self selection of the components of a refuse sample that can be enumerated.

Recently, Qian and Barlaz (1996) enumerated bacteria present on specific constituents of refuse to identify which components of refuse carried refuse-decomposing microorganisms into landfills. Constituents enumerated included grass, leaves, branches, food waste, and mixed refuse, and the sample size was on the order of 10 kg. Thus, it was necessary to modify previously developed inoculum formation techniques. Solid samples were placed in a large (113-l) plastic garbage can which had been wiped with ethanol and purged with sterile argon. A measured volume of filter sterilized anaerobic phosphate buffer (23.7 mM, pH 7.2) was then added to a sample to form a slurry. The sample was then stirred by hand (covered with arm-length gloves). Next, four samples were removed using a 1-l beaker. The contents of a beaker were poured through a second person's hands (covered with gloves) into a sterile, 4-l, argon-purged erlenmeyer flask. Finally, solids caught in the person's hands were squeezed to release excess liquid into the flask. The liquid in this flask served as the inoculum for MPN enumerations. Inocula were serially diluted in phosphate buffer (23.7 mM, pH 7.2). The results of this work are presented in Section 2.3.2.2.

The inoculum formation technique for the 10-kg sample was evaluated in a manner analogous to the work described by Barlaz et al. (1989b). Measured volumes of anaerobically digested sewage sludge were spiked into measured amounts of refuse. The cellulolytic and acetoclastic methanogenic bacteria in refuse, refuse plus sludge, and sludge were then enumerated using the MPN technique. Three spike treatments were evaluated: (1) refuse plus a low level spike of sludge, (2) refuse plus a high level spike of sludge, and (3) refuse plus a low level spike without exposure to refuse particulate matter. For the cellulolytic bacteria, the ratio of the measured-to-expected MPN ranged from 0.72 to 2.22 in the various treatments. This range of measured-to-expected MPNs is not significantly different from 1.0 ($p = 0.05$). The ratio of measured-to-expected MPNs for the acetoclastic methanogenic bacteria was not significantly different from 1.0 for the high level spike (0.23) and the low level spike in the absence of refuse particulate matter (0.94). However, the measured population in the low level spike was significantly lower ($p = 0.05$) than that of the low level spike in the absence of refuse particulate matter. This suggests that sludge methanogenic bacteria attached to refuse particulate matter and this interfered with their enumeration. However, given the dependencies of methanogenic bacteria on soluble substrates and cellulolytic bacteria on insoluble substrates, this conclusion seems unlikely. If attachment were indeed a problem, then low spike recoveries would have been measured for the cellulolytic bacteria as well. The nearly quantitative recovery of cellulolytic and methanogenic bacteria in this experiment suggested that the inoculum formation procedure was repeatable and did not cause a reduction in viable cells. The applicability of the inoculum formation procedure to the other trophic groups which participate in refuse decomposition was assumed.

In summary, researchers have been consistent in their efforts to minimize exposure to oxygen between the time of excavation from a landfill and population enumeration. Inocula have always been formed in an anaerobic buffer solution; however, there is variability in the sample size used for inoculum formation and the precise method of microorganism extraction. As in other environmental samples which include solids, there is no measure of the extent to which attached microorganisms are separated during any inoculum formation technique. Techniques for inoculum formation must be adapted to the objectives of a particular study.

2.3.1.1.2 Microbial Enumeration. Enumerations for total populations have been conducted using AODC, as well as roll tubes and MPN techniques. Adaptation of the AODC procedure to refuse has been presented by Palmisano et al. (1993b). Briefly, between 0.5 and 1 g of refuse were fixed in formaldehyde. Samples were then diluted to a known volume in sodium pyrophosphate and incubated for 5 to 10 minutes. Next, samples were cooled to 0°C and alternately sonicated and cooled. Samples were then stained, filtered, and counted.

Barlaz et al. (1989a) enumerated the total anaerobic population and the subpopulations of cellulolytic, hemicellulolytic, butyrate catabolizing acetogenic, and acetoclastic and hydrogeneotrophic methanogenic bacteria using MPN techniques. The media utilized were presented in the referenced paper with some modifications described later (Wang et al., 1994). The medium for enumeration of the total anaerobic population contained 10 soluble carbon sources (cellobiose, glucose, maltose, xylose, galactose, arabinose, mannose, starch, glycerol, and galacturonic acid), each at a concentration of 2.5 mM. Carbon sources were representative of refuse hydrolysis products. Microbial growth on cellulose was detected by visible disappearance of ball-milled Whatman number 1 filter paper. Xylan from oat spelts (Sigma; St. Louis, MO) was used for enumeration of the hemicellulolytic bacteria. Prior to use, the xylan was soaked in distilled water for 24 hours to remove the soluble and nonsettleable material. Acetogenic bacteria were enumerated based on conversion of butyrate (40 mM) to acetate and hydrogen and subsequent conversion of the hydrogen to methane by a pure culture of *Methanobacterium formicicum*. Methanogen MPN tests were performed with either 80 mM acetate or 202.6 kPa H$_2$/CO$_2$. Tubes were incubated for 30 days, except for the acetogen MPN tubes, which were incubated for 60 days.

Palmisano et al. (1993b) also used a rich medium for enumeration of the total fermentative population. Anaerobic starch and protein-degrading microbes were enumerated on peptone yeast glucose (PYG) agar amended with soluble starch or gelatin, respectively. Zones of starch clearing were detected by the addition of potassium iodide, while proteolytic microbes were detected by the addition of 0.05% napthol blue black in 7% acetic acid. Colonies producing extracellular proteases were surrounded by clear zones.

Roll tubes also have been used for the enumeration and isolation of cellulolytic bacteria from refuse (Westlake and Archer, 1990; Palmisano et al., 1993b). In this procedure, the tubes are inspected visually or with the assistance of a microscope for a zone of clearing indicative of the consumption of insoluble cellulose. Researchers typically use ball-milled cellulose as the cellulose source, although many other alternatives are available. In unpublished work, Barlaz attempted to use strips of Whatman number 1 filter paper in MPN tubes for enumeration of cellulolytics; however, repeatable results were never obtained.

2.3.1.1.3 Measurement of Anaerobic Fungi.   The activity of anaerobic fungi in landfills could be significant, since fungi are often able to attack more recalcitrant, lignicellulosic substrates. There have been two studies in which researchers evaluated the presence of anaerobic fungi in refuse. Barlaz et al. (1989b) inhibited bacteria by the addition of streptomycin sulfate at 130 U/ml and penicillin K at 2000 U/ml. There was no evidence of anaerobic cellulolytic fungi from an inoculum formed from one decomposed refuse sample. Using similar techniques, Theodorou et al. (1989) tested a total of 12 refuse samples from 3 landfills for the presence of anaerobic fungi. Anaerobic fungi were not detected in any sample, although microaerophilic fungi were detected in some of the samples.

2.3.1.1.4 Habitat Simulation.   In developing culture media, an attempt is typically made to simulate the habitat. For example, in studies of the rumen microflora, clarified rumen fluid is added to the medium. In landfill microbiology, adjustments can be made to the media pH and incubation temperature. However, simulation of the organic matrix in a landfill is problematic. Leachate could be used for habitat simulation. One limitation of leachate is that it contains degradable organic carbon. Thus, if the objective of a medium is to grow a specific trophic group, then the presence of uncharacterized carbon in the leachate will interfere with the enumeration assay. A second problem with leachate is that typically there is not a source of constant composition available. As discussed earlier, leachate composition reflects the decomposition state of the refuse. Leachate composition from a landfill will vary as it is affected by the overlying refuse and dilution with stormwater. Leachate from a laboratory-scale lysimeter will vary in composition over time unless the refuse is well decomposed. To date, there are no reports of researchers using leachate in a growth medium.

2.3.1.1.5 Incubation Temperature.   The incubation temperature for an enumeration assay should reflect the temperature of the ecosystem from which the inoculum is recovered. The optimum temperature for methane production in the mesophilic range has been reported as 41°C by Hartz et al. (1982) and

42°C by Pfeffer (1974). Mata-Alvarez and Martinex-Viturtia (1986) found the maximum methane production rate occurred at 42°C, although maximum cumulative methane production occurred at 34 to 38°C. Based on these results, Barlaz et al. (1989a) incubated tubes for MPN tests at 40°C. The refuse used as the inoculum source for these tests were also incubated at 40°C. Palmisano et al. (1993b) incubated samples excavated from the Fresh Kills landfill at both 22 and 37°C. The average temperature of the excavated samples was 29°C, although the range was 10 to 63°C (Suflita et al., 1992).

2.3.1.2 Direct Measures of Microbial Activity

2.3.1.2.1 Enzyme Assays. Techniques used to measure enzyme activities in refuse microorganisms are reviewed here, and the results of such measurements are presented in Section 2.3.2. As in the enumeration of whole cells, enzyme assays require a liquid inoculum. Jones and Grainger (1983) were the first to present a procedure for the measurement of enzyme activity in refuse samples. A 10-g sample of refuse was extracted in 30 ml of a detergent solution of 0.2% Triton X-100 containing 0.75 M $MgSO_4$. After extraction, samples were centrifuged at 10,000 g and the supernatant was used in the enzyme assays.

Esterase activity was measured by the hydrolysis of fluorescein diacetate to fluorescein, the presence of which could be measured by fluorescence spectrophotometry. Protease activity was measured using Azocoll, a proteinaceous substrate which releases a blue dye on hydrolysis. Amylase activity was measured using substrates labeled with a *p*-nitrophenol moiety. In this assay, *p*-nitrophenyl alpha D-maltoheptaoside is converted to *p*-nitrophenyl-maltotriose (PNPG3) plus maltotetraose by alpha-amylase. The PNPG3 is ultimately hydrolyzed to glucose and *p*-nitrophenol, which was measured spectrophotometrically. Cellulase activity was measured based on the degradation of cellulose-azure, which releases a blue dye on hydrolysis. Lipase activity was measured by Jones and Grainger (1983) but not Palmisano et al. (1993a). The assay is based on the titrimetric measurement of fatty acids released from triglycerides. Finally, for a broader characterization, Palmisano et al. (1993a) used the API-ZYM system to screen for the presence of 18 different enzymes.

Westlake and Archer (1990) used a more elaborate technique for measurement of cellulase activity in pure cultures of cellulolytics isolated from decomposing refuse. Cell extracts were obtained by first growing cells in liquid medium and then harvesting them in the late logarithmic phase. After centrifugation, bacterial pellets weighing 1 to 2 g were resuspended followed by sonication. The supernatant, which represented a cell-free whole cell homogenate, was decanted and frozen until use. In addition, a cell-free supernatant fraction was produced by filter sterilization of their original growth medium followed by the addition of ethanol to 70% by volume. The precipitated protein was then harvested by centrifugation, resuspended in buffer, and frozen until use. Carboxymethylcellulase activity was measured by the release of reducing sugars from a 0.5% (w/v) solution of carboxymethylcellulose. Cellobiohydrolase activity was measured by the release of *p*-nitrophenol from *p*-nitrophenylcellobioside. Total cellulase activity was measured as a decrease in turbidity of a 0.025% cellulose suspension using a spectrophotometer at 660 nm.

There are several limitations to the use of enzyme assays as described here. First, all of the work was done with 10-g samples. If the results are to be used for general characterization of the presence or absence of activity, then this sample size may be sufficient. However, 10-g samples cannot be used to extrapolate activity to an entire landfill. In either case, it is important to work with multiple replicate samples to explore spatial variability. Of course, factors such as pH, refuse composition, temperature, and nutrient status will vary spatially. Thus, measurement of spatial variability will reflect variation in conditions throughout a landfill.

A second limitation to enzyme assays is the effectiveness of the cell lysis and washing step. While 0.2% Triton X-100 has been used, there has been no published work comparing its effectiveness to that of other detergents or lysis procedures. In this respect, enzyme activity data are probably best evaluated as relative measures among samples within the same study.

A third limitation is specific to the cellulase assay. Jones and Grainger (1983) evaluated their extraction technique by the addition of a commercial cellulase to both sterile and nonsterile refuse. Upon extraction, the cellulase was fully recovered from the sterile refuse, but only a fixed amount of enzyme activity was measured in the nonsterile refuse regardless of the amount of enzyme added. The authors suggested that the enzyme could have been deactivated by proteases.

2.3.1.2.2 Other Measures of Microbial Activity.   Researchers have used two other techniques to assess the microbial activity of decomposing refuse: (1) estimation of *in situ* methane production, and (2) mineralization of ^{14}C-cellulose. A measure of methane-producing activity may be obtained by placing a sample of freshly excavated refuse in a reactor, sealing it, and sparging it with nitrogen to displace oxygen which entered during reactor loading (Gurijala and Suflita, 1993; Ham et al., 1993). The rate of methane production then can be measured for the excavated sample, with and/or without amendment. The weight of refuse required is flexible as any size reactor can be used. Typically, 5 to 10 kg of wet refuse are used, although Gurijala and Suflita (1993) used 200 to 300 g. As described earlier, methane production from refuse requires the coordinated activity of several trophic groups. Thus, measurement of methane-producing activity provides a measure of the overall activity of the refuse sample as opposed to the activity of a specific organism or enzyme.

Palmisano et al. (1993a) measured the ability of refuse to convert added ^{14}C-cellulose to ^{14}CO$_2$ and ^{14}CH$_4$. Ten-g samples of refuse were placed in bottles to which ^{14}C-cellulose was added. After incubation, mineralization was measured by sparging the headspace of a bottle with nitrogen, forcing headspace gases through a system to trap $^{14}CO_2$ and $^{14}CH_4$. The system included NaOH traps to trap the $^{14}CO_2$, followed by a combustion furnace to oxidize $^{14}CH_4$ to $^{14}CO_2$, and finally another set of traps to dissolve the newly produced $^{14}CO_2$. Trapped radio-labeled gases were then quantified by scintillation counting. Palmisano et al. sampled the system weekly for 4 weeks, although other sampling strategies could be employed. The measure of activity reflected the syntrophic activity of all trophic groups involved in cellulose conversion to methane.

The procedure described above has one important limitation. The author's experience with commercially purchased radio-labeled cellulose has shown that its actual specific activity is as much as ten times greater than the level indicated by the vendor. Thus, where mass balances are involved, it is critical to measure the specific activity independently. The specific activity of commercially purchased ^{14}C-cellulose has been measured successfully by mixing radio-labeled cellulose with powdered cellulose and then ball milling the combination for 5 days. Ball milling serves to dilute the specific activity of the radio-labeled material and to homogenize the small weight of labeled cellulose in a larger mass of unlabeled cellulose. After ball milling, triplicate 0.2-g samples were subjected to the procedure employed for cellulose and lignin analysis (Effland, 1977; Pettersen and Schwandt, 1991). This includes hydrolysis in 72% sulfuric acid followed by a secondary hydrolysis in 3% sulfuric acid. The hydrolyzate will contain hydrolyzed cellulose as glucose. The activity of the hydrolyzate may then be analyzed by scintillation counting for back calculation of the specific activity of the radio-labeled cellulose.

In summary, techniques have been described here for measurement of the population of total and viable cells, enzyme activities, and overall measures of methane production potential. These techniques have been applied to a number of landfill samples; recent research is summarized in the following section.

2.3.2 Landfill Microbiology Research

Much of the initial research on landfill microbiology focused on counts of fecal bacteria and pathogens (Cook et al., 1967; Donnely and Scarpino, 1984; Kinman et al., 1986). Such work is useful for evaluation of potential health problems associated with refuse and landfill leachate but does not elucidate the microbial processes responsible for landfill methanogenesis. Later work focused on the populations required for methanogenic refuse decomposition as well as the activity of hydrolytic enzymes. Some of this work was summarized in a review (Barlaz et al., 1990) and more recent research is presented in this section.

Fielding et al. (1988) characterized seven methanogenic bacterial isolates from landfill samples. Using antigenic fingerprinting to compare the relatedness of the isolates to a reference methanogen culture collection, the isolates were identified as *Methanobacterium formicicum*, *Methanosarcina barkeri*, *Methanobacterium bryantii*, and an unidentified coccus. The optimum pH for all isolates was 7.0, and the optimum temperature was between 37 and 41°C.

Bagnara et al. (1985) were the first to report on the isolation of a cellulolytic microorganism from a sanitary landfill. The isolate was not previously identified and was named *Cellulomonas fermentans*. It is capable of growth under both aerobic and anaerobic conditions but with no apparent advantage to aerobic growth. Its major fermentation products under both aerobic and anaerobic conditions were acetate, formate, and ethanol. The minimum vitamin requirements of the organism were for biotin and thiamine and better growth was exhibited with 0.05% yeast extract. Optimum growth occurred at temperatures between 30 and 37°C and at pH 7.4. Growth was significantly inhibited below pH 7 and above pH 8.

Westlake and Archer (1990) and Westlake et al. (1995) isolated a number of cellulolytic bacteria from refuse samples excavated from landfills in the U.K. and U.S. Cultures were isolated from roll tubes containing ball-milled cellulose. Five isolates were extensively characterized. Two were determined to be clostridia and three were classified in the genus *Eubacterium*. Based on comparisons with known cellulolytic isolates, all isolates were considered to be previously unidentified species (Westlake et al., 1995). The four isolates that were tested were able to grow on ball-milled newsprint at a rate equal to or slightly below their growth rate on ball milled cellulose. All had pH optima between 6.8 and 7.7, while their temperature optima ranged from 37 to 50°C.

Additional work with the isolates of Westlake and Archer was conducted by Cummings and Stewart (1994). They worked with five isolates, although some were different from those initially characterized by Westlake and coworkers. Cummings and Stewart focused on the relative biodegradability of newsprint prior to the application of ink. For three *Eubacterium* isolates, the rate of filter paper degradation exceeded the rate of recycled newsprint degradation which exceeded the rate of virgin newsprint pulp. This result was attributed to the lower lignin concentration in the recycled newsprint (18.5 vs. 24%). Tests with filter paper soaked in ink on one side showed that the ink was a physical impediment to biodegradation but was not toxic. The ink contained carbon black, petroleum distillates, and mineral oil. The optimal temperature was 37°C for four of five isolates, while a fifth exhibited slightly higher activity at 45°C. Gelhaye et al. (1993) isolated five cellulolytic bacteria from a MSW digester, and all were reported to be clostridia. They studied mechanisms of attachment of the isolates and found that they attached to filter paper and coated paper better than to newsprint and corrugated boxes.

Finlay and Fenchel (1991) are the first to report on the presence of anaerobic protozoa in decomposing refuse. They isolated a number of protozoa from excavated refuse samples, including two *Mastigamoeba* species, *Heteromita* sp., *Chilomastix* sp., *Phreatamoeba* sp., and at least one other unidentified heterotrophic flagellate. The most frequently occurring isolate was identified as *Metopus palaeformis* Kahl 1927, and they focused their research on its ecology. The isolate exhibited a high degree of polymorphism through its life cycle and was found to carry methanogenic endosymbionts. Exposure to oxygen inhibited growth but did not cause permanent damage to either the protozoa or the methanogenic symbiont, which remained viable after several days of oxygen exposure. The authors suggested that the dispersal of protozoan cysts containing methanogens is a mechanism for the transport of methanogens through aerobic environments.

Palmisano et al. (1993a) studied hydrolytic enzyme activity and population densities (1993b) in refuse samples excavated primarily from the Fresh Kills landfill in New York as well as a few samples from landfills in Arizona and Florida. Esterase, amylase, and protease activity was present in all 28 samples excavated from the Fresh Kills site. There was no correlation between enzyme activity and moisture content or enzyme activity and the rate of gas production. This conflicts with earlier work by Jones et al. (1983), who reported increasing amylase and protease activity with increasing moisture content.

Cellulase activity was only present in two samples from the Fresh Kills landfill. Three of 8 samples from the Los Reales landfill in Arizona and 1 of 17 samples from the Naples landfill in Florida also had measurable cellulase activity. Cellulose degrading activity also was measured in selected samples by assessing the ability of a sample to convert ^{14}C-cellulose to $^{14}CO_2$ and $^{14}CH_4$. Excavated mixed refuse samples mineralized from <5 to 20% of the added labeled cellulose in 28 days. Samples which contained specific cellulosic substrates such as grass and wood chips mineralized from <5 to 23% of the added cellulose. Less than half of the samples containing specific cellulosic substrates exhibited cellulose mineralization activity. A number of explanations for the overall low evidence of cellulolytic activity were offered, including (1) a limited distribution of cellulolytic microorganisms, (2) the presence of inhibitory compounds, (3) a lack of physical contact between membrane bound cellulases and cellulosic material, (4) nutrient deficiencies, and (5) insufficient moisture.

In parallel work, Palmisano et al. (1993b) measured the total population by AODC, the total viable aerobic and fermentative populations, and the subpopulations of proteolytic, amylolytic, and cellulolytic microorganisms. AODCs were between 1.1 and 5.8×10^{10} cells per dry gram in five samples each from the Fresh Kills and Naples landfills and two samples from the Los Reales landfill. Total aerobes ranged from 0.02×10^6 to 50×10^6 cells per dry gram based on incubations at 22 and 37°C, respectively. Where the populations were significantly different ($p < 0.05$), the counts at 22°C were always higher. Total anaerobes ranged from 0.07×10^6 to 842×10^6 cells per dry gram based on incubations at 22 and 37°C, respectively. Where the populations were significantly different, the counts at 37°C were higher than the counts at 22°C in four of five tests, and there was no significant difference in three samples. The proteolytic population comprised 2 to 15% and 0 to 4% of the total fermentative population at 22 and 37°C, respectively. The amylolytic population comprised 0.4 to 12% and 2 to 4% of the total fermentative population at 22 and 37°C, respectively. Cellulolytic bacteria were not detected in any sample. Methanogenic and sulfate-reducing bacteria were detected in all of the samples tested from the Fresh Kills landfill, although population data were not presented (Gurijala and Suflita, 1993; Suflita et al., 1992).

In summary, there appears to be considerable difficulty with the measurement of cellulolytic bacteria and cellulase activity. Barlaz et al. (1989a) reported relatively low cellulolytic populations in refuse that was actively degrading cellulose. Similarly, Palmisano et al. (1993a, b) detected no cellulolytics in several refuse samples and cellulase activity in only 2 of 28 samples. They suggested that one or more of the enzymes responsible for cellulose hydrolysis was membrane bound and not extracted. This is consistent with work by Westlake and Archer (1990) who reported that cellulose hydrolysis was greater in cell homogenates relative to the extracellular fraction. Interestingly, Cummings and Stewart (1994) found that only three of five cellulolytic bacteria isolated from decomposing refuse could readily degrade filter paper which is nearly pure cellulose. Combined, these data suggest that techniques for the growth of refuse cellulolytic populations and measurement of cellulolytic activity in decomposing refuse are inefficient.

2.3.2.1 Pathogenic Microorganisms

Public health aspects of MSW are reviewed in Chapter 5 and in articles by Pahren (1987) and Fedorak and Rogers (1991). Thus, studies of the presence of pathogens in landfills and leachate are not presented here. For completeness, another aspect of the Fresh Kills landfill study involved evaluation of pathogen survival. Suflita et al. (1992) reported that no viable pathogenic viruses or protozoa were recovered from 54 soiled diapers excavated from the landfill.

2.3.2.2 Microbial Populations on Individual Refuse Constituents

Qian and Barlaz (1996) measured the populations of the trophic groups required for refuse decomposition as they were present on individual components of refuse including grass, leaves, branches, food waste, and fresh refuse. Samples were collected to represent each component as it would enter a landfill. Their data are summarized in Table 2. They concluded that yard waste was the major carrier of the trophic groups required for refuse decomposition.

In summary, increased attention to the landfill ecosystem has resulted in a number of studies on different aspects of landfill microbiology, including the presence and activity of hydrolytic and methanogenic bacteria, the presence of protozoa, and the isolation of refuse microorganisms. These studies will serve as the base from which further research is conducted. The final sections of this chapter present alternate systems for the study of refuse decomposition, relationships between decomposition and gas and leachate production in landfills, and potential avenues of future research.

TABLE 2　Anaerobic Microbial Populations on Refuse and Refuse Components[a]

	Total Anaerobes	Hemicellulolytic	Cellulolytics	Acetogens	Methanogens Acetate	Methanogens (H_2/CO_2)
Grass[b]	9.8	9.2	1.4	1.5	1.6	1.6
Branches	6.5	4.2	2.5	1.3	1.1	0.8
Leaves[b]	6.6	4.3	1.6	4.1	2.7	3.5
Food[b]	9.4	6.0	<-0.4	<-0.1	<-0.1	<-0.1
Refuse[b]	9.1	6.5	0.4	0.2	3.6	4.7

[a] Most probable number, $\log_{10}$ cells per gram dry weight. Data reported as less than a number indicate that no positive tubes were detected. The number reported assumes one positive tube in the first dilution.

[b] Average of two samples.

2.4 SYSTEMS FOR THE STUDY OF REFUSE DECOMPOSITION AND LANDFILL MICROBIOLOGY

A number of systems have been used to study refuse decomposition and landfill microbiology, with each system typically optimized for a specific objective. A number of researchers have worked with samples of refuse from full-scale landfills. Ham et al. (1993) placed samples in 15-cm diameter reactors capable of holding about 10 kg of refuse. The relatively large reactor provides more opportunity to collect a representative sample without selecting sample components because they will not fit in a reactor. While such large reactors may be well suited to measurement of methane-producing activity at the time of sampling, manipulation of a large number of variables in this large of a system would prove difficult. For example, Gurijala and Suflita (1993) also worked with excavated refuse samples but worked with reactors that held only 0.2 to 0.3 kg. The smaller reactors allowed them to manipulate their ecosystem by molybdate addition and to be confident that the added molybdate was in contact with the entire reactor.

Where a laboratory-scale simulation of a landfill is desired, reactors of a variety of sizes have been used. The largest reactors were those used by Pohland and Gould (1986); their system was about 4.2 m long and 1 m in diameter and held about 400 kg of MSW. These reactors are certainly large enough to obtain representative samples of MSW; however, their size makes them difficult to control and manipulate. In addition, they would be of limited use for microbiological studies because it would not be possible to obtain representative subsamples over time. Barlaz et al. (1987) performed work in 210-l steel drums, which have similar advantages and disadvantages to the system used by Pohland and Gould. More recently, Barlaz et al. (1989a) began to work in 2-l reactors with shredded refuse. By shredding refuse, it is possible to obtain representative samples of refuse while working with much smaller quantities. A number of decomposition studies were conducted successfully in 2- to 4-l reactors (Rhew and Barlaz, 1995; Barlaz et al., 1989d). In the initial use of these reactors, multiple replicate reactors were set up and destructively sampled over time to monitor microbial population development during refuse decomposition (Barlaz et al., 1989a).

Whether working with 1 or 400 kg or refuse, the time to the onset of methane production is quite variable. Barlaz et al. (1987, 1992, unpublished work) has shown that the use of a seed of well decomposed refuse will repeatedly initiate refuse decomposition. This has proven ideal for laboratory studies where there is no disadvantage to acceleration of the onset of methane production. Use of a seed will allow decomposition essentially to bypass the anaerobic acid phase.

In studies where the sole objective is to assess ultimate biodegradability, work has been conducted in 125-ml serum bottles using modified Biochemical Methane Potential (BMP) techniques. The BMP test was originally developed by Shelton and Tiedje (1984) to measure the anaerobic biodegradability of a soluble organic chemical. It since has been adapted in various ways for measurement of the biodegradability of solid samples by Bogner (1987), Owens and Chynoweth (1993), and Wang et al. (1994). In the BMP assay, small samples of refuse are added to a serum bottle with liquid media and an inoculum. The samples are dried and shredded and then ground to pass a 1-mm screen before use in the BMP assay. The methane potential of a sample is then measured after incubation for 30 to 60 days. Bogner (1987) modified this procedure somewhat and worked with 25-g samples of excavated refuse which were supplemented with nutrients but not an inoculum. Owens and Chynoweth used the test to measure the methane potential of a number of components of MSW, while Wang et al. used the BMP assay to evaluate the methane potential of excavated refuse samples.

Anaerobic biodegradation is mediated by the activity of bacterial consortia with a wide range of activities and syntrophic interactions. Due to the syntrophic interactions, the study of individual members of anaerobic consortia has always been difficult. A multi-stage continuous culture technique has been applied to the study of landfill microbiology (Coutts et al., 1987; James and Watson-Craik, 1993; Parks and Senior, 1988). In this system, a series of reactor vessels of increasing size are connected and the vessels are operated as a plug flow chemostat. As the reactor size increases, the imposed dilution rate decreases. This selects for faster growing bacteria in the top vessel and slower growing bacteria in the subsequent vessels. This technique is reported to separate habitat domains while still permitting the overlap of activity domains required for syntrophic activity. James and Watson-Craik (1993) used three vessels to segregate hydrolytic/fermentative, acidogenic, and methanogenic activity. The technique was used to study the effect of environmental conditions, including temperature and carboxylic acid concentrations, on methane production.

In summary, there are a number of potential systems available for the study of refuse biodegradation and microbiology. The system to be used must be selected in consideration of specific research objectives.

2.5 FACTORS LIMITING DECOMPOSITION IN LANDFILLS

Numerous researchers have tried to enhance refuse methanogenesis by manipulation of the landfill ecosystem. A number of factors have been shown to influence the onset and rate of methane production including moisture content and moisture flow, pH, particle size, inoculum addition, nutrient concentrations, and temperature, among others. This research has been summarized in a review by Barlaz et al. (1990). The two variables which appear to be most critical in controlling refuse methanogenesis are moisture content and pH.

The moisture content of fresh refuse ranges from 15 to 45% and is typically about 20% on a wet weight basis. While there is no definitive determination as to the minimum or optimal moisture content required for refuse decomposition, 20% is clearly low. Wujcik and Jewell (1980) studied the effect of moisture content on the batch fermentation of wheat straw and dairy manure, compounds analogous in chemical composition to refuse. Methane yields decreased at a moisture content below 70%; the yield at 30% moisture was 22% of the yield at 70% moisture.

Studies on the effect of moisture content on methane production are confounded by refuse pH. Moisture addition stimulates fermentative activity which can lead to an accumulation of carboxylic acids and an acidic pH. Thus, a high moisture content in the absence of pH control can result in a decrease in the time required to reach the anaerobic acid phase. However, in the absence of neutralization, decomposition may remain "stuck" in this phase. In many laboratory-scale studies on the effects of moisture on refuse decomposition, methane yields have been too low for quantification of yield as a function of moisture content.

Emberton (1986) evaluated gas production data for landfills across the U.K. When data from all sites were combined, there was a strong correlation between refuse moisture content and both gas yield and gas production rate, despite the inclusion of confounding factors such as density, refuse age, and refuse composition. Gas production, rather than methane production, was reported, and it was implied that samples producing significant gas had a methane concentration on the order of 50%. Data from two landfills showed large increases in the gas production rate for samples recovered below the water table relative to samples recovered above the water table. Ham et al. (1993) measured methane production in 20 samples excavated from the Fresh Kills landfill in New York. There was a positive correlation between the methane production rate and sample moisture content. Gurijala and Suflita (1993) excavated refuse from the Fresh Kills landfill at the same time as Ham et al. (1993). They showed that addition of moisture enhanced methane production in 67% of the samples tested. Bogner (1990) also reported that moisture stimulated gas production from excavated refuse samples.

A second key factor influencing the rate and onset of methane production is pH. As discussed earlier, the optimum pH for methanogenesis is between 6.8 and 7.4. Segal (1987) found a strong correlation between pH and methane production rate in samples excavated from full-scale landfills. This result was confirmed in samples excavated from the Fresh Kills landfill, where methane production "was largely limited to samples having a circumneutral pH" (Ham et al., 1993; Suflita et al., 1992).

Leachate recycle and neutralization has been shown to enhance the onset and rate of methane production in laboratory-scale tests (Pohland, 1975; Buivid et al., 1981; Barlaz et al., 1987). Given that moisture and pH are reported to be the two most significant factors limiting methane production, the stimulatory effect of leachate recycling and neutralization is logical. Recycling neutralized leachate back through a landfill increases refuse moisture content, substrate availability, and provides a degree of mixing in what may otherwise be an immobilized batch reactor. Neutralization of the leachate provides a means of externally raising the pH of the refuse ecosystem. There is only limited field experience with leachate recycle systems and more is needed to document fully its value in a field-scale situation.

2.5.1 Regulatory Policy Influencing Methane Production

Regulatory policy and philosophy also may limit refuse decomposition. Historically, landfills were not well designed and operated, and this often led to groundwater contamination and the release of contaminated stormwater to lakes and streams. Subsequently, states began to regulate MSW burial, bringing about the engineered sanitary landfill described earlier.

Landfill design standards were developed to minimize the amount of moisture which came in contact with refuse, thus minimizing leachate production. At the time that the philosophy of a dry landfill was adopted, the overriding concern was protection of water quality. The design of landfills to enhance methane production, by allowing the moisture content of fresh refuse to increase by surface water infiltration, was not recognized. More recently, federal landfill regulations have recognized that leachate recycling and neutralization for enhancement of methane production may be advantageous. In addition to energy recovery, enhanced methane production will result in (1) leachate treatment as the soluble organics are converted to CH_4 and CO_2, and (2) more rapid settling of the refuse, which may result in additional capacity for refuse burial. Conversion of CH_4 to CO_2, either during combustion for energy recovery or by a flare, is advantageous as methane is a more damaging greenhouse gas than CO_2 (Tyler, 1991).

2.6 RELATIONSHIP BETWEEN BIOLOGICAL DECOMPOSITION AND LEACHATE AND GAS CHARACTERISTICS

In an anaerobic ecosystem, the composition of the organic matter undergoing biodegradation will be reflected in the ultimate methane yield. Other constituents, either organic or inorganic, may affect the yield by exertion of toxic effects or restriction of biological access to the biodegradable constituents. Earlier in this chapter, it was stated that cellulose and hemicellulose are the major biodegradable constituents in refuse. Recently, a study was conducted to measure the methane yield for the major biodegradable components of refuse (Barlaz et al., 1996). The major components include grass, leaves, branches, food waste, newspaper, corrugated boxes, office paper, and coated paper. The paper components were selected for their prevalence in MSW (USEPA, 1994) and their chemical composition, which should represent the range of biodegradabilities. At one extreme is newsprint, which is a mechanical pulp similar in composition to a tree with the bark removed. At the other extreme is office paper, a chemically produced pulp which has been bleached and nearly completely delignified. Corrugated boxes and coated paper contain

TABLE 3   Measured Methane Yields and Extent of
Decomposition for Each Waste Component

Component	Methane Yield[a] (ml CH_4 at STP per gram dry wt.)
Grass	144.4 (15.5)[b]
Leaves	30.6 (8.6)
Branches	62.6 (13.3)
Food waste	300.7 (10.6)
Office paper	217.3 (15.0)
Coated paper	84.4 (8.1)
Newsprint	74.3 (6.8)
Corrugated boxes	152.3 (6.7)
Seed	25.5 (5.6)

[a] Data have been corrected for background methane production from the seed and are expressed per gram dry weight of test component.

[b] Numbers in parentheses are the standard deviations.

both chemical and mechanical pulp. In addition, coated paper is smooth as a result of the addition of clay and adhesives.

The methane yield of each of the aforementioned components was measured in 2-l reactors in quadruplicate. With the exception of coated paper, each component was shredded. Coated paper was not shredded to avoid increasing the noncoated surface area. Well decomposed refuse excavated from a landfill was used as a seed to initiate decomposition of each component. Reactors were operated to accelerate the rate of decomposition. This included incubation at 40°C, leachate recycling and neutralization, and addition of ammonia and phosphate when concentrations decreased below 100-mg NH_3/l and 10 mg PO_4/l, respectively.

Methane yields for each component are presented in Table 3. All yield data were corrected for background methane production from the seed. The data presented in Table 3 represent ultimate yields in that steps were taken to enhance decomposition. Yields in full-scale landfills would be lower. A limitation of the results presented in Table 3 is that methane potential measurements were not made on mixtures of the components tested; thus, the accuracy of adding the measured methane yields to estimate the yield in a landfill is unknown. By the addition of seed to all reactors and by maintaining macronutrients at concentrations sufficient to support microbial growth, efforts were made to minimize the potential for synergistic effects.

The effects of sulfate on the decomposition process were discussed earlier. A major potential source of sulfate in MSW is scrap wallboard which contains gypsum (calcium sulfate). Analysis of refuse excavated from the Fresh Kills landfill in New York indicated that construction and demolition debris accounted for up to 14% of the buried refuse volume, although the gypsum concentration was not measured (Suflita et al., 1992). Gurijala and Suflita

(1993) showed that by inhibiting sulfate reduction with molybdate, the rate of methane production increased in refuse samples excavated from the Fresh Kills landfill. They also found that various types of paper and textiles excavated from the Fresh Kills landfill had a reservoir of water extractable sulfate. However, extracts of fresh paper did not contain sulfate, suggesting that the absorbant nature of paper allowed for the accumulation of sulfate.

The presence of sulfate will divert some electron flow from methane production to sulfate reduction. Thus, if significant quantities of sulfate are present and distributed throughout the waste, then methane yields could be reduced. In addition, the reduction of sulfate to sulfide results in a mechanism for the precipitation of heavy metals. Theoretically, sulfide could accumulate to inhibitory levels. However, landfill leachate is typically high in iron. Iron would serve as a sulfide sink, given the low solubility of iron sulfides.

There is only limited literature on whether the presence of heavy metals or trace organics inhibits refuse decomposition. Harries et al. (1990) showed that both 100 mg Zn per liter and 5 mg Cu per liter inhibited methane production from the leachate in batch assays. These concentrations are within the ranges detected in leachate. Pohland and Gould (1986) evaluated the addition of a heavy metal sludge containing Zn, Cr, Ni, Cd, Cu, and Fe for its effect on refuse methanogenesis. They established a threshold fraction of sludge that could be added to the refuse without inhibiting refuse decomposition. A few researchers have studied the fate of various organics during refuse decomposition including Deipser and Stegmann (1994), Pohland et al. (1993), and Reinhart and Pohland (1991). However, none of these studies established inhibitory concentrations for trace organics.

Leachate percolating to the bottom of a landfill will either be collected in a leachate collection system or percolate to the water table at a rate governed by the hydraulic gradient and the hydraulic conductivity of the underlying soils. In the absence of a leachate collection system, landfills may be a significant source of groundwater contamination. Chemical and biological mechanisms of leachate attenuation in groundwater were recently reviewed (Christensen et al., 1994). As discussed in Section 2.2.3, it is difficult to relate leachate composition to the state of refuse because leachate samples may represent refuse in several states of decomposition. For the same reason, it is difficult to relate leachate composition to waste characteristics.

2.7 FUTURE RESEARCH DIRECTIONS

Landfills remain a relatively unexplored anaerobic ecosystem in comparison to anaerobic sludge digesters, the rumen, and rice paddies. Thus, there is much opportunity for future research. The major obstacle to much research is the heterogeneity of landfills, making it nearly impossible to conduct studies that provide a picture of an entire landfill. Each researcher must carefully define the objectives of their study and select an appropriate sampling strategy or laboratory simulation to meet their objectives.

One area that has been explored to only a small extent is identification of the dominant microorganisms active during decomposition. The isolation of a number of previously unidentified cellulolytic bacteria was cited above. Only one research team has isolated methanogenic bacteria from landfills. By its nature, a landfill represents a unique anaerobic ecosystem. Environmental conditions and substrate availability change with time and a landfill may never reach a true steady state. Thus, it is not unreasonable to expect the presence of previously uncharacterized microorganisms, with unique strategies for growth and survival.

In addition to isolation using traditional culturing techniques, opportunities exist to use molecular techniques to characterize the landfill ecosystem. Molecular probes could be used to compare molecular characteristics of landfill prokaryotes and archaebacteria to characteristics of known organisms. Such techniques also may help to describe the presence of the viable but nonculturable fraction of the landfill ecosystem.

One aspect of the landfill ecosystem that is beyond the scope of this chapter but worthy of mention is the landfill cover. As explained earlier, the cover consists of a number of layers of soil. The top few centimeters of the cover are likely to be aerobic and offer the opportunity for the aerobic oxidation of methane to carbon dioxide. This is of interest because methane is a greenhouse gas. While the fact that methane oxidation occurs in landfill cover soils has been established (Whalen et al., 1990; Kightley et al., 1995), there is a need to assess the quantitative significance of this process relative to the volumes of methane released through the cover. Such research could lead to the design of landfill covers which maximize methane oxidation.

ACKNOWLEDGMENTS

A number of graduate students and technicians contributed to the work of Barlaz presented herein and their dedication and skill is gratefully acknowledged. These include William Eleazer, III, William Odle, Yu-Sheng Wang, Philip Calvert, and Xindong Qian. Their research was supported by the Air and Energy Engineering Research Laboratory of the U.S. Environmental Protection Agency, Waste Management of North America (now WMX, Inc.), S. C. Johnson & Son, and the National Science Foundation.

REFERENCES

Bagnara, C., R. Toci, C. Gaudin, and J.P. Belaich. 1985. Isolation and characterization of a cellulolytic microorganism, *Cellulomonas fermentans* sp. nov. *Int. J. Syst. Bacteriol.* 35:502–507.

Barlaz, M.A., R.K. Ham, and M.W. Milke. 1987. Gas production parameters in sanitary landfill simulators. *Waste Manage. Res.* 5:27–39.

Barlaz, M.A., D.M. Schaefer, and R.K. Ham. 1989a. Bacterial population development and chemical characteristics of refuse decomposition in a simulated sanitary landfill. *Appl. Environ. Microbiol.* 55:55–65.

Barlaz, M.A., D.M. Schaefer, and R.K. Ham. 1989b. Effects of pre-chilling and sequential washing on the enumeration of microorganisms from refuse, *Appl. Environ. Microbiol.* 55:50–54.

Barlaz, M.A., R.K. Ham, and D.M. Schaefer. 1989c. Inhibition of methane formation from municipal refuse in laboratory scale lysimeters. *Appl. Biochem. Biotechnol.* 20:181–205.

Barlaz, M.A., R.K. Ham, and D.M. Schaefer. 1989d. Mass balance analysis of decomposed refuse in laboratory scale lysimeters. *ASCE J. Environ. Eng.* 115:1088–1102.

Barlaz, M.A., R.K. Ham, and D.M. Schaefer. 1990. Methane production from municipal refuse: a review of enhancement techniques and microbial dynamics, *CRC Crit. Rev. Environ. Control* 19:557–584.

Barlaz, M.A., R.K. Ham, and D.M. Schaefer. 1992. Microbial, chemical and methane production characteristics of anaerobically decomposed refuse with and without leachate recycle. *Waste Manage. Res.* 10:257–267.

Barlaz, M.A., W.E. Eleazer, W.S. Odle, III, X. Qian, and Y.-S. Wang. 1996. Biodegradative analysis of municipal solid waste in laboratory-scale landfills. EPA Final Report. CR-818339. U. S. Environmental Protection Agency, Research Triangle Park, N.C. (submitted for publication).

Bogner, J.E. 1990. Controlled study of landfill biodegradation rates using modified BMP assays. *Waste Manage. Res.* 8:329–352.

Brock, T.D., M.T. Madigan, J.M. Martinko, and J. Parker, 1994. *Biology of Microorganisms,* 7th ed., Prentice Hall, Englewood Cliffs, NJ.

Buivid, M.G., D.L. Wise, M.J. Blanchet, E.C. Remedios, B.M. Jenkins, W.F. Boyd, and J.G. Pacey. 1981. Fuel gas enhancement by controlled landfilling of municipal solid waste. *Resour. Recovery Conserv.* 6:3–20.

Christensen, T.H., P. Kjeldsen, H.J. Albrechsten, G. Heron, P.H. Nielsen, P.L. Bjerg, and P.E. Holm. 1994. Attenuation of landfill leachate pollutants in aquifers. *Crit. Rev. Environ. Sci. Technol.* 24:119–202.

Cook, H.A., D.L. Cromwell, and H.A. Wilson. 1967. Microorganisms in household refuse and seepage water from sanitary landfills. *Proc. W. Va. Acad. Sci.* 39:107–114.

Coutts, D.A.P., E. Senior, and M.T.M. Balba. 1987. Multi-stage chemostat investigations of interspecies interactions in a hexanoate-catabolising microbial association isolated from anoxic landfill. *J. Appl. Bacteriol.* 62:251–260.

Cummings, S.P., and C.S. Stewart. 1994. Newspaper as a substrate for cellulolytic bacteria. *J. Appl. Bacteriol.* 76:196–202.

Daniel, D.E. 1993. *Geotechnical Practise for Waste Disposal.* Chapman & Hall, New York.

Deipser, A., and R. Stegmann. 1994. The origin and fate of volatile trace components in municipal solid waste landfill. *Waste Manage. Res.* 12:128–129.

Donnely, F.A., and P.V. Scarpino. 1984. Isolation, characterization and identification of microorganisms from laboratory and full scale landfills, EPA Project Summary, EPA-600/S2-84-119. U.S. Environmental Protection Agency, Washington, D.C.

Effland, M.J. 1977. Modified procedure to determine acid soluble lignin in wood and pulp. *TAPPI,* 60:143–144.

Emberton, J.R. 1986. The biological and chemical characterization of landfills. Proc. Energy Landfill Gas, Solihull, West Midlands, U.K., Oct. 30–31, 1986.

Emcon Associates. 1980. *Methane Generation and Recovery from Landfills.* Ann Arbor Science, Ann Arbor, MI.

Fedorak, P.M., and R.E. Rogers. 1991. Assessment of the potential health risks associated with the dissemination of micro-organisms from a landfill site. *Waste Manage. Res.* 9:537–563.

Fielding, E.R., D.B. Archer, E.C. de Macario, and A.J.L. Macario. 1988. Isolation and characterization of methanogenic bacteria from landfills. *Appl. Environ. Microbiol.* 54:835–836.

Finlay, B.J., and T. Fenchel. 1991. An anaerobic protozoon, with symbiotic methanogens, living in municipal landfill material. *FEMS Microbiol. Ecol.* 85:169–180.

Fluet, J.E., K. Badu-Tweneboah, and A. Khatami. 1992. A review of geosynthetic liner system technology. *Waste Manage. Res.* 10:47–65.

Francis, A.J., and C.J. Dodge. 1986. Anaerobic dissolution of lead oxide, *Arch. Environ. Contam. Toxicol.* 15:611–616.

Gelhaye, E., L. Benoit, H. Pititdemange, and R. Gay. 1993. Adhesive properties of five mesophilic, cellulolytic clostridia isolated from the same biotope. *FEMS Microbiol. Ecol.* 102:67–73.

Gurijala, K.R., and J.M. Suflita. 1993. Environmental factors influencing methanogenesis from refuse in landfills. *Environ. Sci. Technol.* 27:1176–1195.

Ham, R.K., M.R. Norman, and P.R. Fritschel. 1993. Chemical characterization of Fresh Kills landfill refuse and extracts. *ASCE J. Environ. Eng.,* 119:1176–1181.

Ham, R.K., and T.J. Bookter. 1982. Decomposition of solid waste in test lysimeters. *ASCE J. Environ. Eng.,* 108:1147–1170.

Ham, R.K., K.K. Hekimian, S.L. Katten, W.J. Lockman, R.J. Lofy, D.E. McFaddin, and E.J. Daley. 1979. Recovery, processing, and utilization of gas from sanitary landfills. EPA-600/2-79-001. Municipal Environmental Research Laboratory, U.S. Environmental Protection Agency, Cincinnati, OH.

Harries, C.R., A. Scrivens, J.F. Rees, and R. Sleat. 1990. Initiation of methanogenesis in municipal solid waste. 1. The effect of heavy metals on the initiation of methanogenesis in MSW leachate. *Environ. Technol.* 11:1169–1175.

Hartz, K.E., R.K. Klink, and R.K. Ham. 1982. Temperature effects: methane generation from landfill samples. *ASCE J. Environ. Eng.* 108:629–638.

Iiyama, K., B.A. Stone, and B.J. Macauley. 1994. Compositional changes in compost during composting and growth of *Agaricus bisporus*. *Appl. Environ. Microbiol.* 60:1538–1546.

James, A.G., and I.A. Watson-Craik. 1993. Elucidation of refuse interspecies interaction by use of laboratory models. ETSU B/B2/00148/REP. Energy Technology Support Unit, Department of Trade and Industry, Oxfordshire, U.K.

Jones, K.L., and K.M. Grainger. 1983. The application of enzyme activity measurements to a study of factors affecting protein, starch and cellulose fermentation in a domestic landfill. *Eur. J. Appl. Microbiol. Biotechnol.* 18:181–185.

Jones, K. L., J.F. Rees, and J.M. Grainger. 1983. Methane generation and microbial activity in a domestic refuse landfill site. *Eur. J. Appl. Microbiol. Biotechnol.* 18:242–245.

Kightley, D., D.B. Nedwell, and M. Cooper. 1995. Capacity for methane oxidation in landfill cover soils measured in laboratory-scale soil microcosms. *Appl. Environ. Microbiol.* 61:592–610.

Kinman, R.N., J. Rickabaugh, D. Nutini, and M. Lambert. 1986. Gas Characterization, Microbiological Analysis, and Disposal of Refuse in GRI Landfill Simulators, EPA Project Summary, EPA/600/S2-86/041. U.S. Environmental Protection Agency, Washington, D.C.

Mata-Alvarez, J., and A. Martinex-Viturtia. 1986. Laboratory simulation of municipal solid waste fermentation with leachate recycle. *J. Chem. Tech. Biotechnol.* 36:547–556.

Owens, J.M., and D.P. Chynoweth. 1993. Biochemical methane potential of municipal solid waste (MSW) components. *Water Sci. Technol.* 27:1–14.

Pahren, H.R. 1987. Microorganisms in municipal solid waste and public health implications, *CRC Crit. Rev. Environ. Control* 17:187–228.

Palmisano, A.C., B.S. Schwab, and D.A. Maruscik. 1993a. Hydrolytic enzyme activity in landfilled refuse. *Appl. Microbiol. Biotechnol.* 38:828–832.

Palmisano, A.C., D.A. Maruscik, and B.S. Schwab. 1993b. Enumeration and hydrolytic microorganisms from three sanitary landfills. *J. Gen. Microbiol.* 139:387–391.

Parks, R.J., and E. Senior. 1988. Multistage chemostats and other models for studying anoxic ecosystems, in J.W.T. Wimpenny (Ed.), *CRC Handbook of Laboratory Model Systems for Microbial Ecosystems*, Vol. 1, CRC Press, Boca Raton, FL, p. 51–71.

Pettersen, R.C., and V.H. Schwandt. 1991. Wood sugar analysis by anion chromatography, *J. Wood Chem. Technol.* 11:495–501.

Pfeffer, J.T. 1974.Temperature effects on anaerobic fermentation of domestic refuse. *Biotechnol. Bioeng.* 16:771–787.

Pfeffer, J.T. 1992. *Solid Waste Management Engineering*. Prentice Hall, Englewood Cliffs, NJ.

Pohland, F.G. 1975. *Sanitary Landfill Stabilization with Leachate Recycle and Residual Treatment*. Georgia Institute of Technology, EPA Grant No. R-801397.

Pohland, F.G. and J.P. Gould. 1986. Co-disposal of municipal refuse and industrial waste sludge in landfills. *Water Sci. Technol.* 18:177–192.

Pohland, F.G., W.H. Cross, J.P. Gould, and D.R. Reinhart. 1993. Behavior and assimilation of organic and inorganic priority pollutants codisposed with municipal refuse, EPA/600/R-93/137a, U.S. Environmental Protection Agency Risk Reduction Engineering Laboratory, Cincinnati, OH.

Qian, X., and M.A. Barlaz. 1996. Enumeration of anaerobic refuse decomposing microorganisms on refuse constituents. *Waste Manage. Res.* (in press).

Reinhart, D.R., and F.G. Pohland. 1991. The assimilation of organic hazardous wastes by municipal solid waste landfills. *J. Ind. Microbiol.* 8:193–200.

Rhew, R., and M.A. Barlaz. 1995. The effect of lime stabilized sludge as a cover material on anaerobic refuse decomposition. *ASCE J. Environ. Eng.* 121:499–506.

Robinson, J.A., and J.M. Tiedje. 1984. Competition between sulfate-reducing and methanogenic bacteria for H_2 under resting and growing conditions. *Arch. Microbiol.* 127:26–32.

Segal, J.P., 1987. Testing large landfill sites before construction of gas recovery facilities. *Waste Manage. Res.* 5:123–131.

Shelton, D.R., and J.M. Tiedje. 1984. General method for determining anaerobic biodegradation potential. *Appl. Environ. Microbiol.,* 47:850–857.

Suflita, J.M., C.P. Gerba, R.K. Ham, A.C. Palmisano, W.L. Rathje, and J.A. Robinson. 1992. The world's largest landfill: a multidisciplinary investigation. *Environ. Sci. Technol.* 26:1486–1495.

Tchobanoglous, G., H. Theisen, and S. Vigi. 1993. *Integrated Solid Waste Management*, McGraw Hill, New York.

Theodorou, M.K., C. King-Spooner, and D.E. Beever. 1989. *Presence or Absence of Anaerobic Fungi in Landfill Refuse*. Energy Technology Support Unit (ETSU) Report B 1246, Dept. of Energy, Harwell Laboratory, Oxfordshire, U.K.

Thorneloe, S.A. and J.G. Pacey. 1994. Landfill gas utilization-database of North American projects, Solid Waste Association of North America 17th Landfill Gas Symposium, March 22–24, Long Beach, CA, SWANA, Silver Springs, MD.

Tyler, S.C.1991. The global methane budget, in J.E. Rogers and W.B. Whitman (Eds.), *Microbial Production and Consumption of Greenhouse Gases: Methane, Nitrogen Oxides and Halomethanes*, American Society for Microbiology, Washington, D.C., pp. 7–39.

USEPA. 1988. National Survey of Solid Waste (Municipal) Landfill Facilities. EPA/530-SW88-034. U.S. Environmental Protection Agency, Washington, D.C.

USEPA. 1990. Air Emissions from Municipal Solid Waste Landfills — Background Information for Proposed Standards and Guidelines. Draft, Environmental Impact Statement, Office of Air Quality, U.S. Environmental Protection Agency, Research Triangle Park, NC.

USEPA. 1994a. Characterization of Muncipal Solid Waste in the United States: 1994 Update. EPA530-R-94-042. U.S. Environmental Protection Agency, Washington, D.C.

USEPA. 1994b. International Anthropogenic Methane Emissions: Estimates for 1990. EPA/230-R-93-010. U.S. Environmental Protection Agency, Office of Policy, Planning, and Evaluation, Washington, D.C.

Wang, Y-U, C.S. Byrd, and M.A. Barlaz. 1994. Anaerobic biodegradability of cellulose and hemicellulose in excavated refuse samples. *J. Ind. Microbiol.* 13:147–153.

Westlake, K., and D.B. Archer. 1990. Fundamental Studies on Cellulose Degradation in Landfills. Energy Technology Support Unit (ETSU) Report B 1228, Dept. of Energy, Harwell Laboratory, Oxfordshire, U.K.

Westlake, K., D.B. Archer, and D.R. Boone. 1995. Diversity of cellulolytic bacteria in landfill. *J. Appl. Bacteriol.* 79:73–78.

Whalen, S.C., W.S. Reeburgh, and K.A. Sandbeck. 1990. Rapid methane oxidation in a landfill cover soil. *Appl. Environ. Microbiol.* 56:3405–3411.

Winding, A., S.J. Binnerup, and J. Sorensen. 1994. Viability of indigenous soil bacteria assayed by respiratory activity and growth. *Appl. Environ. Microbiol.* 60:2869–2875.

Wolfe, R.S. 1979. Methanogenesis, in J.R. Quayle (Ed.), *Microbial Biochemistry, International Review of Biochemistry,* Vol. 21. University Park Press, Baltimore, MD.

Wujcik, W.J., and W.J. Jewell. 1980. Dry anaerobic fermentation. *Biotechnol. Bioeng. Symp.* 10:43–65.

Young, L.Y., and A.C. Frazer. 1987. The fate of lignin and lignin derived compounds in anaerobic environments. *Geomicrobiol. J.* 5:261–293.

Zehnder, A.J.B. 1978. Ecology of methane formation, in R. Mitchell (Ed.), *Water Pollution Microbiology,* Vol. 2, John Wiley & Sons, New York, pp. 349–376.

Zeikus, J.G. 1980. Microbial populations in digesters, in D. Stafford (Ed.), *Anaerobic Digestion*, Applied Science Publishers, Englewood, NJ.

3

Anaerobic Digestion of Municipal Solid Wastes

David P. Chynoweth and Pratap Pullammanappallil

CONTENTS

3.1 Introduction 72

3.2 Biomethanogenesis: Microbiology 74
    3.2.1 Depolymerization 74
    3.2.2 Intermediate Reactions...79
    3.2.3 Methanogenesis 80

3.3 Biomethanogenesis: Methods of Study 82
    3.3.1 Culture and Characterization 82
    3.3.2 Immunological and Nucleic Acid Probes 82
    3.3.3 Biomass Measurements..83
    3.3.4 Activity Measurements 84

3.4 Process Modeling...87
    3.4.1 Mass Balances..88
    3.4.2 Kinetics...88
    3.4.3 Mass Transfer Effects 91
    3.4.4 Fluid Dynamics 92

3.5 Operational Parameters 93
    3.5.1 Feedstock Characteristics...93
    3.5.2 Loading...94
    3.5.3 Inoculation...96
    3.5.4 Temperature 96
    3.5.5 Nutrients 97
    3.5.6 Mixing 98
    3.5.7 Pretreatment 98

0-8493-8361-7/96/$0.00+$.50
© 1996 by CRC Press, Inc.

3.6 Performance Parameters 98
    3.6.1 Methane Production and Decomposition of Organic Matter....................98
    3.6.2 Organic Acids, pH, and Alkalinity 99
    3.6.3 Oxidation-Reduction Potential...100
    3.6.4 Reactor Design 100

3.7 Commercialization and Product Use 101

3.8 Conclusion..103

References 104

3.1 INTRODUCTION

Anaerobic digestion of municipal solid waste (MSW) is a controlled process of microbial decomposition where, under anaerobic conditions, a consortium of microorganisms convert organic matter into methane, carbon dioxide, inorganic nutrients, and humus (Chynoweth, 1995). During the process known as *biomethanogenesis,* microorganisms including protozoa, fungi, and bacteria decompose organic matter using carbon dioxide and the methyl group of acetate as electron acceptors in the absence of dioxygen or other electron acceptors. This microbial activity is responsible for carbon recycling in anaerobic environments, including wetlands, rice fields, intestines of animals, aquatic sediments, and manures. Methane is formed from two primary substrates, acetate and hydrogen/carbon dioxide (or formate). In the absence of methanogens to utilize these substrates, organic acids accumulate, which causes a decrease in pH, which ultimately inhibits and stops the fermentation.

Since methane is a significant greenhouse gas, its source fluxes and their potential reduction are of concern (Chynoweth, 1992; USEPA, 1993a,b,c,d). Natural and anthropogenic sources account for 30 and 70%, respectively, of the total methane released per year. Wetlands and animal guts (mainly insects and ruminants) have been identified as major natural sources. The principal anthropogenic sources of methane include fossil fuel processing industries, rice fields, and landfills. Over 80% of the source flux of atmospheric methane is derived from biological activity (Stevens and Engelkemeir, 1988).

Humans have harnessed biomethanogenesis for rapid and controlled decomposition of organic wastes and biomass feedstocks to methane, carbon dioxide, and stabilized residues. In a generalized scheme for anaerobic digestion (Chynoweth, 1987a), feedstock is harvested or collected, coarsely shredded, and placed into a reactor which has an active inoculum of microorganisms required for the methane fermentation. A conventional reactor is mixed, fed once or more per day, heated to a temperature of 35°C, operated at a hydraulic retention time of 15 to 20 days, and loaded at a rate of 1.6 kg volatile solids (VS, organic matter as ash-free dry weight) per cubic meter per day. Under these conditions, a reduction of organic matter of about 50% is achieved, corresponding to a methane yield of 0.3 m^3 per kg VS added. The biogas composition is typically 60% methane and 40% carbon dioxide, with traces of hydrogen sulfide and water vapor. Solid residues may be settled and/or dewatered and used as a soil amendment. The product gas can be used directly or processed to remove carbon dioxide and hydrogen sulfide.

The earliest application of anaerobic digestion was the treatment of domestic and animal wastes (review by McCarty, 1982). Presently, the process is used widely for treatment of municipal sludges and industrial wastes in developed countries (Pohland and Harper, 1985; Switzenbaum, 1991). Developing countries such as China and India still use biogas (anaerobic digester gas) for cooking, lighting, and operation of small engines with the residues applied to fields as a soil amendment (Ke-yun et al., 1988). The process also kills disease-causing organisms resulting in reduced health problems related to fecal contamination (Bendixen, 1994). In the U.S. and other developed countries, commercial application of the process was previously limited primarily to domestic sludges. As energy costs increased and impending depletion of fossil fuels became apparent in the 1970s and early 1980s, the search for renewable alternative fuels resulted in an expanded interest in anaerobic digestion of industrial wastes (Switzenbaum, 1991), MSW (Cecchi et al., 1993), and biomass energy crops (Chynoweth and Isaacson, 1987; Chynoweth, 1995). During this period, several novel high-rate digester designs were commercialized for industrial wastes, predominantly in the food industry. These industries realized the benefits of treating their wastes by a process that eliminates costly aeration requirement, produces less sludge than aerobic processes, and generates a fuel that can offset a portion of the operational energy requirements. Although a few animal waste digesters were placed into operation in developed countries (Pohland and Harper, 1985; Baldwin, 1993, Lusk, 1994), the absence of strict environmental regulations for these wastes and prevailing low energy prices stifled their development. Research on anaerobic digestion of MSW has also expanded since the mid-1970s, resulting in new digester designs for high solids feedstocks (Cecchi et al., 1988). Although small demonstration plants representing these designs were built and operated, low tipping fees and plunging energy prices stifled further commercialization.

Despite relatively low energy prices, interest in renewable energy and related conversion technologies has been revived for several reasons (Chynoweth, 1995). Although the eventual depletion of fossil fuels remains in the background as a long-term incentive for development of sustainable energy, the increased dependency of the U.S. and other countries on foreign imports may be unhealthy for their economies. More urgent incentives to reemphasize renewable energy are related to concerns about global environmental quality. The first concern to emerge was release of toxic compounds and oxides of nitrogen and sulfur resulting from combustion of fossil fuels. These air pollutants contribute globally to health and environmental problems, the most common of which is referred to as *acid rain*. The second and greatest concern, however, is the threat of global warming related to increasing concentrations of carbon dioxide and other upper atmospheric pollutants resulting from anthropogenic activities. Use of renewable biomass (including energy crops and organic wastes) as an energy resource is not only "greener" with respect to most pollutants, but its use represents a closed balanced carbon cycle regarding atmospheric carbon dioxide (Spencer, 1991). It would also mitigate atmospheric carbon dioxide levels through replacement of fossil fuels. A third concern is the recognized need for effective methods for treatment and disposal of large quantities of municipal, industrial, and agricultural organic wastes. These wastes not only are a major threat to environmental quality, but also represent a significant renewable energy resource.This chapter reviews the status of anaerobic digestion as applied to MSW. The principles of the microbiology of biomethanogenesis and their application in the design, operation, and evaluation of the anaerobic digestion process are discussed. Finally, the commercial status of anaerobic digestion is addressed.

3.2 BIOMETHANOGENESIS: MICROBIOLOGY

Biomethanogenesis is the overall anaerobic biological process that involves several general pathways for decomposition of lignocellulosics and other organic complexes and compounds to methane and carbon dioxide (see Chapter 2). Several species of microorganisms are involved in the overall reactions, which include depolymerization, fermentative acidogenesis, acetogenesis, acidogenic back reactions, and methanogenesis (Zinder, 1993). A sustained balanced fermentation requires the activities of microbial consortia for the oxidation of substrates and removal of inhibitory acids, electrons, and hydrogen for the formation of methane. An overall scheme of biomethanogenesis is shown in Figure 1. The principal substrates of methanogenic bacteria are acetate and hydrogen/carbon dioxide (or formate). The relative importance of formate vs. hydrogen/carbon dioxide in the methane fermentation is not well documented because of exchanges which occur between these substrates (Zinder, 1993).

3.2.1 Depolymerization

The first step during the anaerobic digestion of MSW is the depolymerization of polymeric (macromolecular) solid substrates into smaller molecules. We prefer to use the term *depolymerization* for this step, though in the literature this step is frequently referred to as *hydrolysis*. Hydrolysis is only one of many routes to depolymerization. The depolymerization process is mediated by extracellular enzymes secreted by the microorganisms. Depolymerization can be mediated either by hydrolases or lyases, these being the most common modes of enzymatic depolymerization. Hydrolysis reactions are carried out by extracellular enzymes called *hydrolases*.

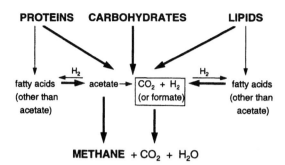

FIGURE 1 Principal reactions of balanced biomethanogenesis. The primary intermediates are acetate, formate, hydrogen, and carbon dioxide. Other acid intermediates are not thought to be formed except from metabolism of odd-numbered carbon skeletons from amino acids and unsaturated fatty acids.

Depending on the type of reaction catalyzed, these hydrolases can be esterases (enzymes that hydrolyze ester bonds), glycosidases (enzymes that hydrolyze glycosidic bonds), or peptidases (enzymes that hydrolyze peptide bonds). For example, lipases hydrolyze the ester bonds of lipids to produce fatty acids and glycerol, glycosidases hydrolyze the polysaccharide component of plant cell walls, and phosphodiesterases hydrolyze the ester bonds of some modified polysaccharides that contain sugars derivatized with phosphoryl, acyl, or alkyl groups (Gander et al., 1993). Lyases, on the other hand, catalyze the nonhydrolytic removal of groups from substrates. For example, pectate lyases depolymerize the pectate component of plant cell wall (Preston et al., 1993).

The products of depolymerization are soluble smaller molecules, and, hence, this step is also known as *solubilization*. The degradable polymeric substrates found in MSW include lignocellulosics, proteins, lipids, and starch (see Chapter 2). Depolymerization of these substrates converts them into a form that can be assimilated into the microbial cell and metabolized. A distinct physiological population, the hydrolytic bacteria, is responsible for depolymerization of these organic polymers and fermentation to products including organic acids, alcohols, and the methanogenic substrates. Usually depolymerization is the slowest step; hence, it is the rate-limiting step in the overall anaerobic digestion process (Eastman and Ferguson, 1981; Noike et al., 1985). In addition, the efficiency of the depolymerization step dictates the ultimate methane yield. In MSW anaerobic digesters, typically only 50% of the organic matter (measured as volatile solids) is depolymerized. The rest of the organic matter remains undegraded due to the inaccessibility of depolymerizing enzymes to sites within the solid matrix and a lack of appropriate organisms that secrete the essential extracellular enzymes.

Lignocellulose is a collective term for the three major components of plant tissue: cellulose, hemicellulose, and lignin. In general, processed MSW contains

40 to 50% cellulose, 12% hemicellulose, 10 to 15% lignin by dry weight (Wang et al., 1994). The sources of these components are paper, paperboard, yard waste, and food waste (see Chapter 1). The cellulose and hemicellulose fractions are biodegradable and make up over 90% of the biochemical methane potential of MSW. However, not all the hemicellulose and cellulose are bioavailable in anaerobic digestion (Wang et al., 1994). This is primarily due to the architecture of the plant cell wall (Tsao, 1984). Cellulose is a linear homopolymer of several thousand D-glucose units linked by beta-1,4 glucosidic bonds. From a degradation viewpoint these linkages are easy to hydrolyze; however, it is not the primary linkage but rather the tertiary and secondary structures of cellulosic materials that make it difficult to degrade. Cellulose is contained in the plant cell wall and surrounding it is a heavily lignified material called the *middle lamella*. The middle lamella contains lignin and hemicellulose in a proportion of approximately 7:3. Since enzymatic hydrolysis of the glucosidic bonds is mediated after the adsorption of the enzymes to cellulose, the middle lamella presents a barrier to the movement of the enzymes. Hence, the diffusion and penetration by the enzymes is extremely difficult and slow. Moreover, the phenolic groups in lignin might even be inhibitory to the enzymes.

Hydrolases degrade cellulose to yield a soluble disaccharide, cellobiose, which on further hydrolysis results in D-glucose. Cellulose-hydrolyzing enzymes from different microbial species have been isolated and investigated. Several anaerobic bacteria degrade cellulose, including *Bacteriodes succinogenes, Clostridium lochhadii, Clostridium cellobioporus, Ruminococcus flavefaciens, Ruminococcus albus, Butyrivibrio fibrisolvens, Clostridium thermocellum, Clostridium stercorarium,* and *Micromonospora bispora* (Tsao, 1984; Gilkes et al., 1991; Linden and Shiang, 1991). The properties of cellulases are very similar, although they are produced by different organisms. The cellulolytic enzyme system is composed of endoglucanases, exoglucanases, and glucosidases (Eriksson et al., 1990). Endoglucanases randomly split glucosidic linkages, exoglucanases split off either cellobiose or glucose from the nonreducing end of the cellulose, and glucosidases hydrolyze cellobiose and other water soluble cellodextrins to glucose (Eriksson et al., 1990). In addition to these enzymes, some anaerobic bacteria produce phosphorylases to mediate cellulose degradation (Ljungdahl and Eriksson, 1985). Two types of phosphorylases have been distinguished, namely those specific for cellobiose and those utilizing higher cellodextrins. The products of depolymerization by these enzymes are glucose-1-phosphate and glucose. A detailed review on biological degradation of cellulose has recently been published by Béguin and Aubert (1994).

Among the anaerobic bacteria, the cellulase system of *Clostridium thermocellum* has been widely studied (Ljungdahl et al., 1983; Lamed and Bayer, 1988; Wu and Demain, 1988). *C. thermocellum* produces a low molecular weight, water insoluble, yellow affinity substance which promotes the binding of cellulolytic enzymes to cellulose (Ljungdahl et al., 1983). It also has been shown that in anaerobes the cellulolytic system is produced as an aggregate called the cellulosome. Cellulolytic aggregates are produced by *C. thermocellum, B. succinogenes,* and *R. albus* (Eriksson et al., 1990). These cellulolytic aggregates help to bind the cellulose to the cell and subsequently to hydrolyze the cellulose. The synthesis of cellulases is regulated by induction and catabolite repression mechanisms (Linden and Shiang, 1991; Eriksson et al., 1990). Cellulase synthesis is induced by soluble derivatives of cellulose or low molecular weight carbohydrates such as cellobiose, lactose, and sophorose. Enzyme synthesis is repressed by the presence of glucose or readily metabolizable sugars. Most of our knowledge on the cellulose-degrading anaerobic bacteria comes from studies conducted on rumen microorganisms. The predominant hydrolytic microorganisms in rumen differ from anaerobic digestion systems (Rivard et al., 1991). In rumen systems, *Ruminococcus* and *Bacteriodes* predominate, whereas in anaerobic digesters fed with MSW the predominant hydrolytic bacteria are *Clostridia* (Rivard et al., 1991). In addition to bacteria and protozoa, the rumen population also may include anaerobic fungi which play an important role in cellulose degradation (Eriksson et al., 1990). The presence of anaerobic fungi in anaerobic digesters, however, has not yet been shown.

Rivard et al. (1991) found that cellulase activity in anaerobic digesters was low relative to other cellulolytic environments such as the rumen, suggesting that MSW digesters were operating under less than optimal cellulase activity. Lagerkvist and Chen (1993) showed that the addition of a commercially available fungal cellulase preparation that has endoglucanase, exoglucanase, and cellobiohydrolase as its principal activities increased the conversion of cellulose by almost 50% under both acidogenic and methanogenic conditions. These results indicate that in some systems the availability of cellulolytic enzymes may be rate limiting. However, results obtained by Rintala and Ahring (1994) suggest otherwise. They studied the effect of additions of both active and inactive enzymes such as xylanase, lipase, protease, and a mixture of these on thermophilic digestion of household sorted waste. They concluded that enzyme addition to household solid waste increased neither the rate of methanogenesis nor the methane yield.

Hemicelluloses are composed of both linear and branched heteropolymers of D-xylose, L-arabinose, D-mannose, D-glucose, D-galactose, and D-glucuronic acid (Eriksson et al., 1990). In most cases, these sugars are linked by 1,4-beta linkage, except for galactose-based hemicelluloses where the linkages are of the type 1,3-beta. The products of depolymerization are the monomers that make up the hemicellulose polymer. Hemicelluloses are more readily degraded than cellulose by anaerobic microbes because they are simpler structures, less branched, and less tightly bound within the lignocellulosic complex. Even though depolymerization of hemicellulose is an easy process, the complexity of hemicellulolytic enzyme systems far exceeds that of the cellulolytic enzymes, as hemicellulose is composed of more varieties of monomers (Tsao, 1984). The predominant rumen bacteria that degrade hemicelluloses are *Bacteriodes ruminicola, B. fibrisolvens, R. flavefaciens,* and *R. albus.*

Pectins represent an important group of hemicelluloses in young plant tissues, berries, and fruit. The quantity of pectin in lignified plant materials varies between 1 and 4%. Pectins consist of rhamnose and unbranched chains of beta-1,4 glycosidically linked to D-galacturonic acid units which are either partly or completely esterified with methanol and also contain other constituents such as galactose and arabinose (Senior and Balba, 1990). Depolymerization of pectin is mediated by a range of lyases and hydrolases that includes esterases and glycosidases. Preston et al. (1993) studied the activity of pectate lyase secreted by the anaerobic bacteria, *Clostridium populeti*, isolated from a wood digester. They found that the organism synthesizes a nonrandom trimer-generating pectate lyase with a combination of endolytic and exolytic depolymerizing mechanisms. Several *Clostridium* species have been identified as pectinolytic (Ng and Vaughn, 1963). Pectinolytic bacteria isolated from rumen fluid include *Bacteriodes rumenicola* and *Streptococcus bovis* (Wojciechowicz and Tomerska, 1971).

Lignin is a highly branched, constitutionally undefined aromatic polymer composed of phenylpropane subunits that are randomly linked by a variety of carbon-carbon and ether bonds. Depolymerization of lignin produces homocyclic aromatic compounds, and anaerobic bacteria are able to utilize these aromatic compounds in several anaerobic energy-yielding processes such as anoxygenic photosynthesis, denitrification, sulfate reduction, fermentation, and methanogenesis (Elder and Kelly, 1994). Though anaerobic bacteria are capable of degrading the monomeric units that make up the lignin molecule, it is doubtful whether lignin can be depolymerized to these monomers under conditions that prevail in anaerobic digesters (Odier and Artaud, 1992). Recent investigations suggest that anaerobic bacteria may be involved in some transformations of lignified plant tissues in the rumen (Colberg, 1988).

Starch is readily biodegradable; two types of starch molecules exist in nature, amylose and amylopectin. Amylose is an unbranched polymer and consists of glucose residues in 1,4-alpha linkages, while amylopectin is the branched form and has one 1,6-alpha linkage per 30 1,4-alpha linkages. The 1,6-alpha linkages initiate the side chains. An enzyme system for complete hydrolysis of starch is made up of multiple enzymes that act synergistically. Three main types of enzymes are often found in such a system (Tsao, 1984): (1) alpha-amylases, which cleave the 1,4-alpha bonds; (2) beta-amylases, which break off maltose from the nonreducing end of amylose and amylopectin; and (3) glucoamylases, which cleave off individual glucose units from an amylolytic chain at its nonreducing end. Some of the microbes found in anaerobic digesters capable of degrading starch are *Streptococcus bovis, Bacteriodes amylophilus, Selenomonas ruminatium, Succinomonas amylolytica, B. ruminocola,* and a number of *Lactobacillus* species (Tsao, 1984).

Proteins are hydrolyzed by proteolytic enzymes to peptides, amino acids, ammonia, and carbon dioxide. In anaerobic digesters, proteins serve as a source of carbon and energy for bacterial growth and as a source of nitrogen from ammonia released during hydrolysis. It has been shown that a specialized group of anaerobic bacteria such as the proteolytic clostridia (e.g., *Clostridium perfringens, C. bifermentans, C. histolyticum, and C. sporogenes*) is responsible for protein degradation in digesters (McInerney, 1988).

Fatty acids are the main constituents of lipid fractions found in wastes. This class of lipids contains both highly hydrophobic and highly hydrophilic regions. Simple lipids (fats and oils) are esters of fatty acids with a glycerol backbone. These are also called *triglycerides* because of three fatty acids being linked to the glycerol molecule. Complex lipids (e.g., phospholipids, glycolipids) contain additional constituents such as phosphate, nitrogen, and sulfur or small hydrophilic carbon compounds such as sugars, ethanolamine, serine, or choline. Plant tissues contain between 1 and 25% (dry weight) of lipids (Senior and Balba, 1990). Esterase-mediated enzymatic hydrolysis of lipids results in the release of saturated and unsaturated long chain fatty acids together with glycerol. Glycerol is easily assimilated and metabolized by the bacteria. The long chain fatty acids undergo an intracellular beta-oxidation mediated by a variety of enzymes, resulting in the production of organic acids such as acetic acid and propionic acid along with hydrogen. Some of the lipid solubilizing anaerobes found in MSW anaerobic digesters are *Anaerovibrio lipolytica* and *Syntrophomonas wolfei* (Cecchi and Mata-Alvarez, 1991).

3.2.2 Intermediate Reactions

Products of depolymerization reactions are converted to fermentation products, which in a balanced methanogenic environment are primarily hydrogen and carbon dioxide, formate and acetate. Odd-numbered carbon skeletons may produce some other fatty acids such as propionate. Under conditions of electron or hydrogen accumulation (e.g., when methanogenesis is inhibited) numerous other fermentation products may be formed, including propionate, butyrate, lactate, succinate, and alcohols, as a way of removing electrons or hydrogen. Organisms that convert fermentation products, such as propionate, butyrate, lactate, and ethanol, generally exhibit obligate proton-reducing metabolism. Dihydrogen is produced as a fermentation product, and this reaction is obligately dependent on hydrogen removal by methanogenic or other hydrogen-using bacteria (Zinder, 1993). This mechanism is commonly referred to as interspecies hydrogen transfer. The organisms are referred to as syntrophs and may be obligately syntrophic, as in the case of *Syntrophomonas wolfei* and *Syntrophobacter wolinii*, or facultative, as with many other syntrophs. The fact that hydrogen strictly inhibits methanogenic fermentation of benzoate (Kamagata et al., 1992) suggests the importance of that mechanism in the metabolism of toxic organic compounds. In addition, there are organisms present which form acetate and other C-3 or higher volatile acids via back reactions with dihydrogen and carbon dioxide (Chynoweth and Mah, 1971; Boone and Mah, 1987).

Recently, the potential significance of interspecies acetate transfer in biomethanogenesis has been suggested (Zinder, 1993); accumulation of acetate in the fermentation, in addition to causing a reduction in pH, may inhibit acetogenic reactions by thermodynamic mechanisms. This specific inhibition has been demonstrated for acetogenesis from propionate (Stamms et al., 1992) and butyrate (Beaty and McInerney, 1989). Zinder and Koch (1984) and Ahring (1995) have shown that interspecies hydrogen transfer may be important in acetate metabolism in thermophilic digesters where the conversion of acetate may involve oxidation to hydrogen and carbon dioxide as an intermediate step instead of the more common direct aceticlastic conversion of the methyl group to methane. This was shown to be the predominant mechanism in thermophilic digesters where the acetate pool was less than 1 mM and *Methanothrix* was the predominant acetate-using methanogen (Ahring, 1995).

Longer chain organic acids (propionate and above) accumulate when the rate of hydrolytic and fermentative activity exceeds the rate of acetogenic conversion of fermentation intermediates to acetate and dihydrogen. This is usually because methanogens cannot consume dihydrogen at the rate at which it is produced. In a balanced fermentation, acids other than acetate and formate are only formed from odd-numbered carbon skeletons (e.g., from decomposition of amino acids and unsaturated fatty acids). This theory is based on the following evidence. First, in the presence of a high hydrogen consumption, the thermodynamics favor production of hydrogen rather than fermentation products as a method of electron disposal (Zinder, 1993). Second, a defined consortium, including a cellulolytic, acetolytic, and hydrogenolytic bacteria, did not produce acids other than acetate (Weimer and Zeikus, 1977; Wolin and Miller, 1982). Third, organisms exist in the methane fermentation which form acetate and other acids by back reactions (Chynoweth and Mah, 1971; Boone and Mah, 1987). And, fourth, balanced anaerobic digesters have a limited capacity to utilize propionic acids (Pullammanappallil et al., 1994).

The overall fermentation, therefore, is dependent upon a delicate balance between activities of bacteria that form organic acids, carbon dioxide, and dihydrogen (or formate) and methanogenic bacteria which utilize these substrates. Overproduction of electrons and acetate, resulting from overfeeding or under-utilization of electrons and acetate by methanogenic bacteria, results in the accumulation of fermentation products to inhibitory levels. This results in the cessation of decomposition. Because organic acids, including propionate and larger, are not major intermediates in a balanced methane fermentation, they are not metabolized after formation during imbalance until a population of bacteria capable of their metabolism can develop. The growth rate of propionate-utilizing bacteria is slow in comparison to other organisms, so this process can often require weeks.

3.2.3 Methanogenesis

The methanogenic bacteria are such a unique group of organisms that they have been placed into a new evolutionary domain (separate from eucaryotic plants and animals and procaryotic bacteria) referred to as *Archaea* (Woese et al., 1990), formerly known as *Archaebacteria* (Woese, 1987). Archaea also includes other species of extreme halophilic and thermophilic bacteria. Placement of methanogens into a separate taxonomic group acknowledges the difference in their genetic makeup from that of other living organisms including bacteria. These genotypic differences are reflected in numerous phenotypic characteristics unique to this group, including metabolism, coenzymes, and cell membrane lipids (Zinder, 1993). Methanogenic substrates include acetate, methanol, dihydrogen/carbon dioxide, formate, carbon monoxide, methylamines, methyl mercaptans, and reduced metals (Figure 2). Methanogens have unique coenzymes for electron transfer: CoM and F_{420} (Zinder, 1993). The fluorescence property of F_{420} has been used to locate methanogenic colonies and enumerate methanogens in mixed culture (Peck and Archer, 1989; Peck and Chynoweth, 1992). The property of possessing ether-linked instead of ester-linked lipids in their cell membranes (deRosa et al., 1986) has been applied as a method to distinguish methanogens from other microorganisms in the environment (Hendrick and White, 1993).

Hydrogen: $\quad 4H_2 + CO_2 \rightarrow CH_4 + 2H_2O$

Acetate: $\quad CH_3COOH \rightarrow CH_4 + CO_2$

Formate: $\quad 4HCOOH \rightarrow CH_4 + 3CO_2 + 2H_2O$

Methanol: $\quad 4CH_3OH \rightarrow 3CH_4 + CO_2 + 2H_2O$

Carbon monoxide: $\quad 4CO + 2H_2O \rightarrow CH_4 + 3H_2CO_3$

Trimethylamine: $\quad 4(CH_3)_3N + 6H_2O \rightarrow 9CH_4 + 3CO_2 + 4NH_3$

Dimethylamine: $\quad 2(CH_3)_2NH + 2H_2O \rightarrow 3CH_4 + CO_2 + 2NH_3$

Monomethylamine: $\quad 4(CH_3)NH_2 + 2H_2O \rightarrow 3CH_4 + CO_2 + 4NH_3$

Methyl mercaptans: $\quad 2(CH_3)_2S + 3H_2O \rightarrow 3CH_4 + CO_2 + H_2S$

Metals: $\quad 4Me^0 + 8H^+ + CO_2 \rightarrow 4Me^{++} + CH_4 + 2H_2O$

FIGURE 2 Principal methanogenic reactions.

3.3 BIOMETHANOGENESIS: METHODS OF STUDY

The microbial ecology of biomethanogenesis is difficult to study because the microorganisms are fastidious, slow-growing anaerobes, and many species will not grow in pure culture (Chynoweth, 1987a). When grown in pure culture, isolates may produce fermentation products different from those produced in the presence of hydrogen- and acetate-metabolizing bacteria that occur in their natural environment (Chynoweth and Mah, 1971; Weimer and Zeikus, 1977; Wolin and Miller, 1982). Each anaerobic environment may differ in the types of bacteria involved in methanogenesis, depending on differing factors such as substrate, retention time, temperature, pH, and fluctuations in other environmental parameters.

3.3.1 Culture and Characterization

A number of techniques have been developed to study methanogenesis. Initially, anaerobic culturing techniques using roll tubes (Hungate, 1967) and anaerobic chambers (Balch et al., 1979) facilitated isolation and culture of methanogens and other anaerobes. Most probable number (MPN) techniques have been used extensively to estimate physiological groups (Zhang and Noike, 1994; Ahring, 1995). The fluorescence of coenzyme F_{420} facilitates recognition of some species of methanogenic bacteria from nonmethanogenic bacteria in viable mounts and colonies in roll tubes and plates (Lee et al., 1987; Vogels et al., 1980; Zinder and Koch, 1984). Certain of these traditional enumeration techniques have been used to estimate variations in populations of organisms in digesters during startup and two-stage operation (Anderson et al., 1994) and as a function of hydraulic retention time (Zhang and Noike, 1994).

3.3.2 Immunological and Nucleic Acid Probes

Immunological (Brigmon et al., 1994; Macario and de Macario, 1983; Robinson and Erdos, 1985) and nucleic acid (Kane et al., 1991) probes have been developed that permit rapid identification of pure cultures and determination of population dynamics of species in environmental systems. The tedium of isolation, identification, and characterization of bacteria in environmental samples has made it technically unfeasible to characterize individual populations in microbial communities. With the discovery of molecular probes and their applications, it is now possible to develop probe libraries to target individual strains as well as larger physiological groups.

Recognizing that the methanogenic bacteria are genetically unique and of major significance in anaerobic environments, Macario and de Macario (1983) developed immunological probes to facilitate their rapid identification. This method originally was applied to identification and verification of new isolates. Subsequently, both monoclonal and polyclonal antibody probes have been used to estimate populations of methanogens in a variety of anaerobic environments, including anaerobic digesters (Macario and de Macario, 1988; Macario et al., 1989, 1991). Antibody-based probes also have been used in conjunction with gold labeling to identify methanogens in electron micrographs of bacteria from digesters (Robinson and Erdos, 1985). Brigmon et al. (1994) used antibody probes with the ELISA technique to evaluate the significance of a new cellulolytic isolate in wood-fed digesters. *Clostridium aldrichii* was shown to be a predominant organism in this digester for a period of greater than one year. These results were confirmed by direct characterization and identification by conventional culture methods (Yang et al., 1990). Immunological probes have been applied to different digester designs and operating conditions and used to characterize instability.

Ribosomal sequences have documented the dramatic genetic uniqueness of the Archaea (Woese et al., 1990). In the process of this discovery, the 16S ribosomal RNA of many methanogenic cultures was sequenced, providing the basis for creating probes useful for identification of isolates and populations of these organisms in various environments. Raskin et al. (1994b) developed eight oligonucleotide probes (Table 1) which are complementary to conserved tracts of 16S rRNA from phylogenetically defined groups of methanogens. These probes were utilized to characterize methanogens in acetate-fed chemostats and digesters fed sewage sludge or MSW. Figure 3 shows the similarities and differences of methanogenic populations from the waste digesters based on the percentage of total 16S rRNA. Note that methanobacteriales were predominant in the mesophilic and thermophilic MSW digesters, but not in the sewage sludge digesters. Methanosarcinae were abundant in the mesophilic MSW and sewage sludge digesters but not the thermophilic MSW digester. All four commercial sewage sludge digesters exhibited similar methanogenic probe profiles, but they were different from the MSW digester profiles. In particular, the methanosaeta were more prevalent in the sewage sludge digesters. Raskin et al. (1994a) also reported high numbers of sulfate-reducing bacteria in digesters receiving MSW. These results are just the beginning of a new era of using molecular probes for characterizing digesters. As the libraries of probes expand, this tool will facilitate our understanding of microbial populations and their influence on digester performance and startup.

3.3.3 Biomass Measurements

Several general and specific measurements have been made to estimate numbers or mass of microorganisms in anaerobic digestion. These have been reviewed by Switzenbaum et al. (1990) and Peck and Archer (1989). Well established methods such as DNA, protein, dehydrogenase, and ATP continue to be used as measures of total biomass or activity (Switzenbaum, 1991).

The use of bacterial lipids deserves special mention. Membrane lipids not only serve as markers of specific species and groups of bacteria, but also can be used to distinguish between broad groups. Phospholipids can be used as a measure of living bacteria as their turnover is rapid and can be used to distinguish between living and dead cells (Hendrick and White, 1993). Since

**TABLE 1   Oligonucleotide Probes and Relevant Characteristics
of Each Target Group**

Probe	Target Group	Relevant Characteristics
MC1109	Methanococcales	Most use H_2–CO_2 and formate
MB310	Methanobacteriales	Most use H_2–CO_2 and formate; some use H_2–CO_2 and formate
MG1200	*Methanogenium* relatives	Most use H_2–CO_2 and formate
MSMX860	Methanosarcinaceae	Most use acetate
MS1414	*Methanosarcina* plus relatives	All use methanol and methylamines; some use acetate and H_2–CO_2
MS821	*Methanosarcina*	Use acetate and other substrates (H_2–CO_2, methanol, and methylamines); generally have high minimum threshold, K, and μ max values for acetate
MX825	*Methanosaeta*	Use only acetate; generally have low minimum threshold, K, and μ max values for acetate
ARC915	Archaea	Broad spectrum domain probe

Source: From Raskin, L. et al. 1994a. *Appl. Environ. Microbiol.* 60:1241–1248. With permission from American Society for Microbiology.

bacteria in the Archaea domain contain ether-linked rather than ester-linked lipids (Tornebene and Langworthy, 1979), measurement of ether-linked lipids can be related to the total biomass of methanogenic bacteria. Lipid profiles also have been shown to be useful in describing the state of microbial populations of digester under different operating conditions and states of performance (Hendrick and White, 1993; Hendrick et al., 1991a,b).

3.3.4 Activity Measurements

Activity measurements still remain significant tools for examining the role of different physiological groups. Overall decomposition is best measured by the rate of methane production (Kelly and Chynoweth, 1979; Owen et al., 1979). The biochemical methane potential (BMP) assay has been refined and used for measurement of ultimate yields from substrates, toxicity, and conversion kinetics (Chynoweth et al., 1993). Sorensen and Ahring (1993) have refined the biochemical methane potential assay for measurement of specific methanogenic activity of anaerobic digester organisms in response to various substrates and conditions. The rate of methane production and methane gas content are well established measures of digester performance. Chynoweth et al. (1994) made use of this measure as a reliable parameter for real-time, on-line control of digesters under conditions of underloading, overloading, and feed inhibition.

Several gas component analyses have been used to evaluate microbial activity and possible control of digester performance, including hydrogen and

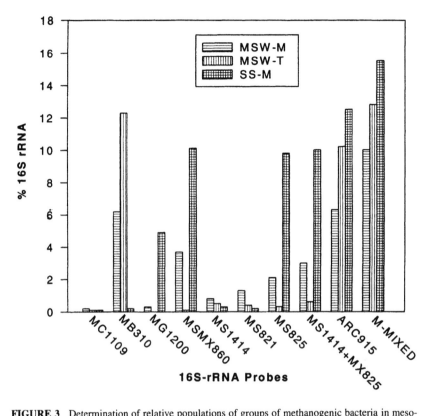

FIGURE 3 Determination of relative populations of groups of methanogenic bacteria in mesophilic (M) or thermophilic (T) digesters receiving municipal solid waste (MSW) or
sewage sludge (SS). Probes are identified in Table 1. (From Raskin, L. et al. 1994a.
Appl. Environ. Microbiol. 60:1241–1248. With permission from American Society
for Microbiology.)

carbon monoxide. Hydrogen was thought to have potential as a performance
parameter because of its link to interspecies hydrogen transfer and regulation
of the methane fermentation. Mosey and Fernandes (1989) showed that hydrogen content in the gas phase responds quickly to disturbances. Laboratory
measurements of gaseous hydrogen, however, are often extremely difficult due
to its low concentrations (<50 ppm under normal operating conditions). Collins
and Paskins (1987) used an exhaled gas monitor to measure hydrogen, but
found that the measurements were contaminated by the presence of H_2S.
Hickey et al. (1987) have reported the presence of carbon monoxide in the
gas phase of anaerobic digesters; carbon monoxide evolved from acetate during
methanogenesis and showed a strong correlation to acetate concentration in
the liquid phase.

Fluorometry has been used successfully to measure reduced NAD(P) and
factor F_{420} as indicators of total and methanogenic activities, respectively

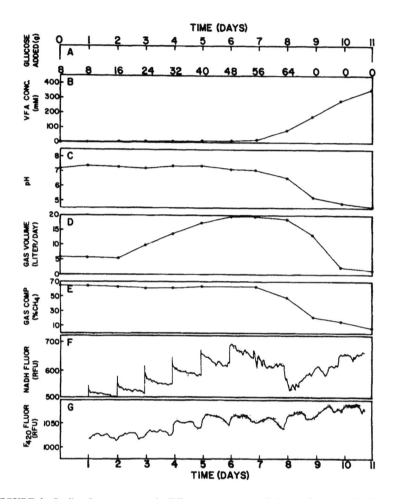

FIGURE 4 On-line fluorescence and off-line measurements of glucose-fed anaerobic digester subject to overloading. The loading was increased from 8 g glucose per day on day 1 to 64 g on day 8 while other components of the feed remained constant. No feed was added on days 9, 10, and 11. Samples for volatile fatty acids, gas composition, and pH were taken immediately before feeding each day. (From Peck, M.W., and D.P. Chynoweth. 1990. *Biotechnol. Lett.* 12:17–22. With permission from Chapman and Hall.)

(Armiger et al., 1986; Peck and Chynoweth, 1990, 1992). Fluorometric measurements showed that reduced NAD(P) responded prior to other measures of digester failure (caused by overloading) including pH, volatile fatty acids, and methane production (Figure 4). Factor F_{420} measured by this technique was not significantly responsive.

Other activity measurements utilize enzyme activity, microcalorimetery, or stable isotope fractionation. Adney et al. (1989) adapted available assays to measure the activities of hydrolytic enzymes in the supernatant of anaerobic digesters and also developed a detergent extraction protocol for releasing particle-bound enzymes. Assays on digester supernatant detected the activities of alpha-amylase and a protease. However, employing the detergent extraction protocol, activities of cellulase, alpha-glucosidase, beta-glucosidase, endoglucanase, exoglucanase, protease and alpha-amylase were recovered from sludge particulates (Adney et al., 1989; Rivard et al., 1994). This indicated that most of the enzymes are strongly bound to the insoluble substrates in the digester. This would be expected because depolymerization is brought about after the enzymes have adsorbed on the surface of particles.

Microcalorimetry is a potentially useful measure of microbial activity and has been applied on a limited basis to digesters (Switzenbaum et al., 1990). The heat energy required to maintain a preset culture temperature is related to overall microbial activity. The measurement can respond to over- and underloads, as well as toxicity. It also would be useful in estimating the influence of reaction heat on digester temperatures which can be significant in high-solids designs characteristic of anaerobic digesters. Stable isotope fractionation using high resolution mass spectrometry has served as a useful tool to distinguish biogenic from abiogenic methane as well as general substrate sources and specific metabolic precursors of biogenic methane (Krzycki et al., 1987; Oremland, 1988; Tyler, 1991). Other useful but more traditional parameters related to microbial activity and discussed below include production and turnover rates of intermediate fermentation products and related parameters such as pH, buffering capacity, and oxidation-reduction potential.

3.4 PROCESS MODELING

Numerous models have been developed to provide a theoretical understanding of microbial populations and their interactions with the physical and chemical environment. Models use mathematical expressions to describe the interactions between various microbial populations involved in the process, including substrate utilization rates, microbial growth rates, product formation rates, and physico-chemical equilibrium relationships. Anaerobic digestion models can be used for optimizing process design and operation and for process control. From the earlier sections, it is clear that the interactions between the microbial consortia in an anaerobic digester are complex. Models that attempt to describe these interactions also tend to be complex and may not be operationally useful. Hence, they often are simplified and contain only the most important overall interactions. For example, models that have been developed for solid waste or plant biomass digestion incorporate the three major steps mediating the degradation process — namely, depolymerization and solubilization, acidogenesis, and methanogenesis (Smith et al., 1988; Mata-Alvarez et al., 1992; Negri et al., 1993).

3.4.1 Mass Balances

The starting point for developing a process model for anaerobic digestion is developing mass balance equations which account for the changes in concentration of the substrates, microbial populations, and products during the course of the digestion process. These balances can be performed for each component involved in the degradation process. Since most anaerobic digesters operate within a narrow temperature range (mesophilic or thermophilic), temperature fluctuations can be ignored. A generalized mass balance applied to an elemental volume in a digester can be written as follows:

Accumulation of mass = (mass in by bulk flow and diffusion)
      − (mass out by bulk flow and diffusion)
      + (generation of new mass)
      − (interphase transfer)

For a batch digester, neglecting diffusional terms and interphase mass transfer and assuming it is well stirred, the mass balance around the digester can be written as:

Accumulation of mass = generation of mass

For continuously operated, well-stirred digesters and making similar assumptions as before, the mass balance equation is:

Accumulation of mass = (mass in by bulk flow)
      − (mass out by bulk flow)
      + (generation of mass)

Commonly, MSW digesters are packed with the waste, and the waste is allowed to degrade. During this period, leachate along with inoculum and nutrients may be recirculated continuously or periodically through the solid bed. Hence, in reality, the operation of MSW digesters could be as batch reactors, with respect to the solid waste, and semi-batch or continuous reactors, with respect to leachate/liquid phase. These equations, when expressed mathematically, become differential equations with the component concentration being the dependent variable. A set of differential equations is obtained when mass balances are written for all the components including the microbial populations. These equations are nonlinear and hence are solved numerically for the component concentrations.

3.4.2 Kinetics

The kinetics of anaerobic digestion has been the subject of a critical review by Pavlostathis and Giraldo-Gomez (1991). The discussion here will be limited to kinetic models that have been applied to anaerobic digestion of MSW.

Several rate expressions have been proposed to describe the generation term in the mass balance equation. Depending on the component in the mass balance, the generation term can be a rate of utilization of the component, the net generation rate (i.e., difference between component generation rate and component utilization rate), or a microbial growth rate. The rate of component formation is assumed to be proportional to the rate of utilization of the substrate from which it is formed; the proportionality constant is the product yield factor. Some of the frequently used substrate utilization rate expressions are listed in Table 2.

A rate expression for microbial growth (X) can be written in terms of the corresponding substrate (S) utilization rate as follows:

$$\frac{dX}{dt} = Y\left(-\frac{dS}{dt}\right)$$

where Y is the growth yield coefficient expressed as milligrams of biomass per milligrams of substrate. McCarty (1971) proposed a method to calculate the growth yield coefficient (Y) from a knowledge of the energetics of the conversion reactions. This method was later used by Labib et al. (1993) to estimate Y in a dynamic model for anaerobic butyrate-degrading consortia. Experimentally determined values of Y also have been reported (Gujer and Zehnder, 1983).

A simple rate expression is obtained if the depolymerization of the MSW and solubilized complex organic molecules is assumed to follow first order kinetics. There is only one parameter that needs to be estimated to describe this process. First order depolymerization parameters for some biopolymers found in waste and for the organic fraction of MSW can be found in Mata-Alverez et al. (1992), Pfeffer (1974), Negri et al. (1993), and Gujer and Zehnder (1983). Cecchi et al. (1990) deduced a semi-empirical equation taking into account the possible diffusion of the enzymes to the particulate matter (degradation rate is proportional to the square root of the substrate). This is called the *step diffusional model* (Table 2).

Another model that has been developed to describe the solubilization and depolymerization of solid waste is known as the *shrinking core model* (Negri et al., 1993). The rate of depolymerization is assumed to be proportional to the fluid-solid surface and the enzyme concentration, which is considered to be proportional to the concentration of acidogenic microorganisms. It is assumed that the acidogenic microorganisms solubilize the solids and metabolize them to acids. Assuming a spherical geometry for the particle, equations were derived for the shrinking size of the particle and disappearance of the insoluble substrate as function of the size (Table 2).

Depolymerization and solubilization reactions of anaerobic digestion are followed by acidogenesis and methanogenesis. Rate expressions usually used to describe these steps follow Monod (1949) kinetics (Table 2). Representative

TABLE 2   Kinetic Models Used in Anaerobic Digestion of Municipal Solid Waste

Models	Substrate Utilization Expressions	Nomenclature
First Order[a]	$-\dfrac{dS}{dt} = kS$	S = substrate concentration k = first order rate constant
Shrinking Core[b]	$-\dfrac{d\phi}{dt} = \beta X_h$ $-\dfrac{dS}{dt} = 3\beta S_0 \phi^2 X_h$	S_o = initial substrate concentration X_h = hydrolytic bacteria concentration ϕ = dimensionless particle radius ($= r/R_p$) r = radius of particle at time t R_p = initial radius of particle β = heterogenous hydrolysis rate (m^{3}d^{-1}kg^{-1})
Step Diffusion[c]	$-\dfrac{dS}{dt} = \left[v_{max}^2 - k(S_0 - S) \right]^{1/2}$	v_{max} = maximum substrate degradation rate k = kinetic constant
Monod[d]	$-\dfrac{dS}{dt} = \dfrac{\mu_{max} S}{Y(K_S + S)} X$	u_{max} = maximum specific growth rate (d^{-1}) K_s = half-saturation constant (mg/l) Y = growth yield coefficient (milligrams biomass per milligrams substrate) X = concentration of bacteria (mg/l)
Inhibition[e]	$-\dfrac{dS}{dt} = \dfrac{\mu_{max}}{Y\left(1 + \dfrac{K_S}{S} + \dfrac{S}{K_I}\right)} X$	K_I = inhibition parameter

[a] Pfeffer (1974)

[b] Negri et al. (1993)

[c] Cecchi et al. (1990)

[d] Monod (1949)

[e] Andrews and Graef (1971)

K_s valves for acidogenesis and methanogenesis are 200 and 15 mg COD l^{-1}, respectively. Valves for u_{max} are 2.0 and 0.4 (Pohland, 1992). Other expressions also have been used to model anaerobic digestion of soluble feedstock (Grau et al., 1975; Contois, 1959; Chen and Hashimoto, 1979). To account for mass transfer limitations, the effluent substrate concentration is expressed as a function of the influent substrate concentration in these models.

It has been observed that high concentrations of substrate can inhibit microbial growth. Andrews and Graef (1971) proposed a model for microbial growth that is inhibited by high concentrations of substrate (see Table 2) by modifying the Monod model. This is relevant because the growth of methanogenic bacteria is known to be inhibited by high concentrations of volatile organic acids (>10,000 mg/l). Low pH can also inhibit growth. Costello et al. (1991) proposed the following expression for the pH inhibition factor:

TABLE 3   Concentrations of Inhibitors of Anaerobic Digestion

Inhibitor	Concentrations (mg/l)
Phenol[a]	2400
Heavy metals[b]	
Zn (II)	160
Fe (III)	1750
Cd (II)	180
Cu (II)	170
Cr (III)	450
Cr (VI)	530
Nickel[a]	250
NH_4^+-N[a]	6000
Calcium[c]	2500–8000
Magnesium[c]	1000–3000
Potassium[c]	2500–12000
Sodium[c]	3500–8000
Sulphide (S^-)[a]	600

[a] Parkin et al. (1983)

[b] Mosey and Hughes (1975)

[c] WPCF (1987)

$$pH_{inhf} = \left[\frac{pH - pH_{LL}}{pH_{UL} - pH_{LL}} \right]^m \quad \begin{array}{l} = 0, \text{ if } pH = pH_{LL} \\ = 1, \text{ if } pH = pH_{UL} \end{array}$$

where m = power coefficient = 3, pH_{LL} = lower limit of pH, and pH_{UL} = upper limit of pH. Biomethanogenesis is sensitive to inhibitors, including alternate electron acceptors (oxygen, nitrate, and sulfate), sulfides, heavy metals, halogenated hydrocarbons, volatile organic acids, ammonia, and cations (Table 3) (McCarty, 1964; WPCF, 1987; Speece, 1987b; Lin, 1993; Mueller and Steiner, 1992). The toxic effect of an inhibitory compound depends upon its concentration and the ability of the bacteria to acclimate to its effects. The inhibitory concentration depends upon different variables, including pH, hydraulic retention time, temperature, and the ratio of the toxic substance concentration to the bacterial mass concentration. Antagonistic and synergistic effects are also common. Methanogenic populations usually are influenced by dramatic changes in their environment but can be acclimated to otherwise toxic concentrations of many compounds.

3.4.3 Mass Transfer Effects

The models described above, except for the step diffusional model, ignore mass transfer resistances to the transport of enzymes and products to and from the substrates. The effect of mass transfer limitations is reflected by the estimated values of the half saturation constant (K_s). As mass transfer limitations become severe, K_s would increase (Pohland, 1992). Mass transfer resistances can occur during transport of enzymes from the microorganism to the bulk fluid and subsequent transport from bulk fluid to the substrate surface. Similarly, mass transfer resistances can occur to products when being transported from substrate surface to bulk fluid. Jain et al. (1992) concluded that mass transfer resistance at the particle scale controlled the rate of hydrolysis. The mass transfer resistance was attributed to the transport of the depolymerizing enzymes from the site of production within cells to the bulk solution where enzymes were accessible to solid substrates. In general, with respect to mass transfer and cellulolytic activity, attachment of cells to the substrate has the advantages of concentration of cellulolytic enzymes and less competition for hydrolysis products with other organisms. A disadvantage is that inhibitory products are not removed as easily from the site of formation.

The transport of hydrolyzing enzyme was simulated as follows:

$$\frac{\partial C_b}{\partial t} = K_1 a \left(C_m - C_b \right)$$

where $K_1 a$ = mass transfer coefficient (per hour)
         C_b = concentration of enzyme in the bulk liquid
         C_m = concentration of enzyme in a film surrounding the microbe

The gas-liquid mass transfer rate can play an important role in determining the ionic equilibrium in the liquid. Carbon dioxide, which is produced during anaerobic digestion, is soluble in the liquid and the dissolved carbon dioxide can affect the pH. The pH, in turn, along with the dissociation constant determine the ratio of ionized to nonionized species of organic compounds in the digester. Gas transfer rates can be calculated using Henry's Law (Andrews and Graef, 1971).

3.4.4 Fluid Dynamics

Since the medium for the transport of nutrients and inoculum in the solid bed is the leachate, factors that may play an important role in determining the efficiency of the process, rate of conversion, distribution of inoculum and nutrients, leachate quality and flow rate are mass transfer resistances between regions of stagnant and flowing fluid and the hydraulic conductivity of the bed. Stagnant zones can develop behind irregular particles and cause pressure losses. Though this phenomenon has not been modeled in high solids anaerobic digesters, models have been proposed in the soils and fiber literature. The hydraulic conductivity of the bed is influenced by the particle size and volume fraction of the solids in the bed. Jackson and James (1986) found that the hydraulic conductivity of various fibrous materials could be expressed as a function of these two parameters. Al-Yousfi and Pohland (1993) demonstrated that applying the concept of probability-based entropy to model hydraulic conductivity was more effective in simulating flow through highly channelled systems. Incorporating these phenomenon into process models would increase the complexity of the models, thus rendering them less amenable to simulation using a personal computer.

3.5 OPERATIONAL PARAMETERS

3.5.1 Feedstock Characteristics

Production and composition of MSW vary from site to site and are influenced by various factors, including region, climate, extent of recycling, use of in-sink disposals, collection frequency, season, and cultural practices. In considering MSW as a feedstock for anaerobic digestion, it is important to understand the feed characteristics (Table 4) (see also Chapter 1). In general, the major components are paper and putrescible fractions (yard and food wastes) which typically comprise over 50% of the wet weight. In developing countries, the organic fraction may be higher because of the effective removal of recyclables by scavengers. In countries such as Denmark and Switzerland, the organic fraction is concentrated by source separation (Christensen and Hjort-Gregersen, 1994; Edelmann and Engeli, 1993).

TABLE 4   Organic Composition of Municipal Refuse

Constituent	Dry Weight[a] (%)	Volatile Solids[b] (%)	Dry Weight[c] (%)	Conversion[d] (%)
Volatile solids	78.6	—	73	58
Cellulose	51.2	40	32.9	75
Hemicellulose	11.9	—	5.2	94
Protein	4.2	5.6	9.6	10
Lignin	15.2	27.3	12.5	17
Lipids	—	6	5.9	66
Starch + soluble sugars	0.5	3.3	—	—
Pectin	<3	—	—	—
Soluble sugars	0.35	—	—	—

[a] Barlaz et al. (1990)

[b] Ten Brummeler (1991)

[c] Peres et al. (1992)

[d] Peres et al. (1992); 35°C; 20-day hydraulic retention time

Organic matter can be digested effectively as unsorted or sorted MSW (Pfeffer, 1974; Pfeffer and Kahn, 1976); however, the degree of separation of organics influences materials handling and the marketability of the residues as a soil amendment (Haskoning, 1994). The trend in sorting is toward source separation of the organic and nonorganic fractions (IEA, 1994). This not only facilitates sorting of recyclables from the nonorganic fraction, but also results in digester feedstocks (and thus residues) that are relatively free of undesired components such as plastics, metals, glass, and heavy metals.

From a microbiological point of view, the organic fraction of MSW has a high solids content (~50%) and limiting nitrogen content (C/N > 30). The principal organic components are cellulose and hemicellulose. The biochemical methane potential of several MSW components was determined to compare the potential extent and rate of their conversion to methane (Owens and Chynoweth, 1993). These data, summarized in Table 5, indicate that a typical conversion efficiency of MSW to methane is 50%, corresponding to a methane yield of 0.2 m^3/kg VS. The highest methane yields were observed for office paper and food packaging. The lowest methane yield was observed for newspaper, and ink did not influence its biodegradability. As expected, the biodegradability of different types of yard wastes was quite variable. These data provide a basis for predicting potential methane production from wastes with known composition. This assay can also be used to determine remaining biodegradability in digester or landfill samples. The undegraded fraction undoubtedly consists of lignin and cellulose that is tightly complexed with lignin which is refractory to anaerobic metabolism.

3.5.2 Loading

The most meaningful parameter for describing the feed rate is loading rate, typically expressed as weight of organic matter (VS or chemical oxygen demand, COD) per culture or bed volume of reactor per day (Fannin and Biljetina, 1987). This parameter accurately describes the reactor size needed for a particular feed rate. Other parameters, such as solids concentration and retention time, are misleading and do not provide a valid basis for comparison of digester costs. Solids concentration does not indicate the feed rate, and the hydraulic retention time (HRT) varies significantly with solids concentration.

Independent of influencing digester size, solids concentration has a significant effect on digester design and performance and materials handling (Chynoweth, 1987b). Feeds with low concentrations of suspended solids (<2%) can be digested in high rate, attached-film reactors such as the upflow sludge anaerobic blanket (UASB), anaerobic filter, and fluidized bed (Switzenbaum, 1991). These reactors retain high concentrations of attached microorganisms and permit low hydraulic retention times without organism washout. Attached-film digesters may be employed for treatment of leachate from landfills (Blakey et al., 1992; Suidan et al., 1993) or soluble feed streams from multistage MSW digesters (Ghosh, 1984; Hack and Brinkman, 1992). Designs for medium solids (2 to 10%) require either high retention times (>15 days) or some mechanism of retaining suspended solids such as solids recycle or concentration of solids within the reactor (Chynoweth et al., 1987; Chynoweth, 1987b; Fannin and Biljetina, 1987). This results in a ratio of solids retention time (SRT) to HRT of greater than 1 for increased retention of solids and

TABLE 5 Estimates of Ultimate Methane Yield (Y$_u$)
and First Order Rate Constants (k) for
MSW Components

Sample	Y$_u$ (m^3/kg volatile solids)	k (d^{-1})
Controls		
Cellulose (a)	0.37	0.09
Cellulose (b)	0.37	0.13
Organic fraction of MSW		
Sumter[a]	0.22	0.075
Levy[a]	0.2	0.073
Yard waste		
Blend	0.14	0.067
Grass	0.21	0.13
Leaves	0.120	0.084
Branches	0.13	0.035
Paper		
Office	0.37	0.14
Corrugated	0.278	0.058
News (no ink)	0.084	0.084
News (with ink)	0.100	0.069
Cellophane	0.35	0.10
Food board (uncoated)	0.34	0.12
Food board (coated)	0.33	0.14
Milk carton	0.32	0.087
Wax paper	0.34	0.83

[a] Organic fraction of MSW prepared by hand/mechanical separations for composting conversion plants.

Source: From Owens, J.M., and D.P. Chynoweth. 1993. *Water Sci. Technol.* 27:1–14. With permission of Pergamon Press.

microorganisms. Designs with higher reactor solids concentrations (>10%) are considered for high solids feeds. A number of advantages of high solids designs (often referred to as dry digestion) include higher potential loading rates, lower heat energy requirements, and less water as a waste product (Wujcik and Jewell, 1980). Rivard et al. (1994) compared the discrete cellulase activities of low-solids and high-solids digesters fed with MSW. It was found that the activities of the discrete cellulases — namely, those of beta-D-glucosidase, endoglucanase, and exoglucanase, on a per gram of sludge basis — were similar. In a high-solids system, however, the overall level of cellulolytic enzyme activity per unit volume is far greater than that of a low-solids system. This greater level of cellulolytic activity per unit volume allows for increased organic loading rates without loss of conversion efficiency (Rivard et al., 1994). High-solids designs have a unique set of advantages and limitations with respect to materials handling related to feed addition, mixing, and effluent removal.

Hydraulic retention time also may have a significant influence on performance, in particular for continuous-flow, stirred-tank reactor (CSTR) digesters. In such digesters, solids and microorganisms wash out, resulting in reduced performance (Fannin and Biljetina, 1987). The critical washout HRT for organisms is related to their growth rate, which is different for different physiological groups in the digester fermentation (Pohland, 1992). In a biomethanogenic reactor, differential washout of critical groups of microorganisms leads to imbalance and deterioration of performance (O'Rourke, 1968). Vinzant et al. (1990) studied the effects of retention time on cellulose degradation in MSW-fed CSTR anaerobic digesters. They reported that substantial cellulose degradation occurs only at retention times greater than 20 days.

3.5.3 Inoculation

The quality and quantity of inoculum is critical to the performance and stability of biomethanogenesis during anaerobic digestion. In conventional CSTR digesters, the inoculum-to-feed ratio (on a VS basis) is typically greater than 10. In designs where washout of critical organisms is a concern, suspended solids in the effluent may be settled and recycled (Fannin and Biljetina, 1987). With batch and plugflow designs, inoculum must be added with the feed. For studies using the BMP assay, a critical inoculum-to-feed ratio of 2 was determined experimentally (Chynoweth et al., 1993). Lower inoculum levels may lead to imbalance due to the more rapid growth rate of acid-forming bacteria (compared to methanogens) and depression of pH. Depending upon the buffering capacity (alkalinity), a digester may be able to recover from low inoculum rates. In some leach-bed designs (Chynoweth et al., 1991), acids formed during startup may be removed via leachate to a started-up combined-phase or methane-phase digester for conversion. Inocula also may become imbalanced when exposed to toxic substances or environmental stress factors (e.g., abnormal temperature) for which they are not acclimated.

3.5.4 Temperature

Biological methanogenesis has been reported at temperatures ranging from 2°C (in marine sediments) to over 100°C (in geothermal areas) (Zinder, 1993). Most applications of this fermentation have been performed under ambient (20 to 25°C), mesophilic (30 to 40°C), or thermophilic (50 to 60°C) temperatures. In general, the overall process kinetics double for every 10°C increase in operating temperature (O'Rourke, 1968) up to some critical temperature (about 60°C) above which a rapid dropoff in microbial activity occurs (Harmon et al., 1993). The populations operating in the thermophilic range are genetically unique (Zinder, 1993), do not survive at lower temperatures, and are more sensitive to temperature fluctuations outside of their optimum range.

Bacteria in thermophilic digesters exhibit some differences compared to those in mesophilic digesters. For example, at thermophilic temperatures, acetate is oxidized by a two-step mechanism (syntrophic acetate oxidation to hydrogen and carbon dioxide followed by formation of methane) when the acetate is less than 1 mM (Zinder and Koch, 1984; Ahring, 1995). At higher concentrations of acetate and mesophilic temperatures, the principal acetoclastic mechanism is direct conversion of the methyl group to methane. Also, ammonia is more toxic in thermophilic digesters due to a higher proportion of free ammonia (Angelidaki and Ahring, 1994). Although thermophilic digesters are thought to have higher energy requirements (Srivastava, 1987), heat losses can be minimized through effective insulation and use of heat exchangers to reduce system heat losses. Most commercial anaerobic digesters are operated at mesophilic or ambient temperatures (Pohland and Harper, 1985). Thermophilic operation is practiced under rare circumstances when the reduced reactor size justifies the higher energy requirements and added effort to ensure stable performance.

3.5.5 Nutrients

Nitrogen and phosphorus are major nutrients required for anaerobic digestion. An average empirical formula for an anaerobic bacterium is $C_5H_7O_2NP_{0.06}$ (Speece, 1987a). Thus, the nitrogen and phosphorus requirements for cell growth are 12 and 2%, respectively, of the VS converted to cell biomass (about 10% of the total VS converted); this would be equivalent to 1.2 and 0.024% of the biodegradable VS, respectively, for nitrogen and phosphorus.

Previous studies have identified critical feedstock C/N ratios of 25 for the organic fraction of MSW (OFMSW) (Kayhanian and Hardy, 1994) above which nitrogen was limiting. In fact, nutrient limitations are better related to concentrations; for example, a value of 700 mg/l was recently reported for the optimum NH_3-N concentration in high solids anaerobic digestion of OFMSW (Kayhanian, 1994). Nutrients also may be concentrated by certain design and operating practices to concentrate nutrients extracted from the feedstock (Chynoweth et al., 1987; Chen et al., 1990; Chynoweth et al., 1992; O'Keefe et al., 1993). Ammonia is also an important contributor to the buffering capacity in digesters (WPCF, 1987) but may be toxic to the process at concentrations above 3000 mg/l. In high-solids digesters, ammonia toxicity was exhibited from feeds that had normal C/N ratios because ammonia was concentrated in the supernatant during digestion (Jewell et al., 1993; Kayhanian and Hardy, 1994).

Other nutrients needed in intermediate concentrations include sodium, potassium, calcium, magnesium, chlorine, and sulfur. Requirements for several micronutrients have been identified, including iron, copper, manganese, zinc, molybdenum, nickel, and vanadium (Speece, 1987a). Available forms of these nutrients may be limiting because of their ease of precipitation and removal by reactions with phosphate and sulfide. Limitations of these micronutrients have been demonstrated in reactors where the analytical procedures failed to distinguish between available and sequestered forms (Jewell et al., 1993).

3.5.6 Mixing

Mixing is traditionally thought to be required for optimized digestion to enhance interaction between cells and substrates and to remove inhibitory metabolic products from the cells (WPCF, 1987). Conventional digesters include mixing which is accomplished by mechanical stirring, liquid recycle, or gas recycle. One design mixes inoculum with feed and then follows with plug flow operation (Six and De Baere, 1992). Leach-bed designs do not mix the solids but use leachate to wet, inoculate, and remove inhibitory organic acids during startup (Ghosh, 1984; Chynoweth et al., 1991). The energy requirement related to mixing is relevant to anaerobic digestion, as mixing can require as much as 14% of the methane energy product in conventional low-solids designs. Rivard et al. (1995a) found that energy requirements for digesters operated at 30% solids were similar to those for low-solids systems (2 to 10%) because higher mixing speeds were needed by the latter to prevent settling of solids and scum formation.

3.5.7 Pretreatment

Most anaerobic digestion systems employ some type of pretreatment to enhance materials handling and microbial conversion. Solid and liquid fractions often are separated by settling, flotation, or pressing. The resulting fractions can be more optimally digested in attached-film and high solids digesters. Removal of nonbiodegradable waste stream components or recyclable materials facilitates materials handling and enhances the value of the digester residues as a compost. Particle size reduction is usually practiced to facilitate materials handling and increase the surface area available for microbial activity. In the case of the OFMSW, shredding enhances mixing of solids, liquid, and microorganisms but does not significantly increase the surface area of the waste particles until particle sizes of less than a few millimeters (the thickness of paper and other waste components) are obtained.

   Other pretreatment processes involving heat, chemical, irradiation, and enzymatic operations have been studied for their ability to enhance the extent and rate of conversion (Tsao, 1987). In general, most methods substantially improve the rate and, to a limited degree, the efficiency of conversion, but the benefits do not justify the added cost to the conversion system.

3.6 PERFORMANCE PARAMETERS

3.6.1 Methane Production and Decomposition of Organic Matter

Total gas and methane production are directly related to the extent and rate of conversion of organic matter which is expressed as VS or COD. Methane yield is preferred over gas yield because pH changes in the reactor can cause changes in release or uptake of carbon dioxide that are unrelated to biodegradation. Use of VS permits calculation of a materials balance between the feed, effluent solids, and gas. Use of COD allows for calculation of an oxidation-reduction balance between the feeds and products. In the context of materials balances, the reduction in organic matter may be calculated as a reduction in VS or COD. A typical methane yield for the organic fraction of MSW is 0.2 m³/kg VS which corresponds to a VS reduction of 50%. As shown in Table 5, methane yields as high as 0.37 l/g VS have been reported for office paper and as low as 0.08 l/g VS for newspaper.

The methane production rate is a measure of process kinetics and is often determined as volume of methane per volume of reactor per day (vvd). This parameter is the product of loading rate (kg/m³/day) and methane yield (m³/kg VS added). Values for MSW digestion have been reported in the range of 1 to 5 vvd. Methane content of the gas is also a good indicator of stability. Under normal circumstances, this value is a function of the H/C ratio of the biodegradable fraction and is normally in the range of 50 to 60% for MSW (Owens and Chynoweth, 1993). Since methanogenic activity is the key factor leading to imbalance, a reduction of methane gas content is an important performance parameter and has been employed as an on-line control parameter (Chynoweth et al., 1994).

The BMP assay (Owen et al., 1979; Owens and Chynoweth, 1993; Chynoweth et al., 1993) is useful for estimating the ultimate methane yield and relative conversion rates of feed samples, specific feed components, and remaining biodegradable matter in process residues. This assay also may be used to determine toxicity of feed components. In general, the test is conducted with miniature digesters (200 ml) which are optimized for conversion in terms of inoculum, feed concentration, nutrients, and buffer. These miniature batch digesters are incubated until no further gas production is observed. Measurements include gas production and composition of influent and effluent organic matter.

3.6.2 Organic Acids, pH, and Alkalinity

Organic acids, pH, and alkalinity are related parameters that influence digester performance (McCarty, 1964; WPCF, 1987). Under conditions of overloading and the presence of inhibitors, methanogenic activity cannot remove hydrogen and organic acids as quickly as they are produced. The result is accumulation of acids, depletion of buffer, and depression of pH. If uncorrected via pH control and reduction in feeding, the pH will drop to levels which stop the fermentation. Independent of pH, extremely high volatile acid levels (>10,000 mg/l) also inhibit performance. The major alkalis contributing to alkalinity are ammonia and bicarbonate. A normal, healthy volatile acid-to-alkalinity ratio is 0.1. Increases to ratios of 0.5 indicate the onset of failure, and a ratio of 1.0 or greater is associated with total failure. The most common chemicals for pH control are lime and sodium bicarbonate. Lime produces calcium bicarbonate up to the point of solubility of 1000 mg/l. Sodium bicarbonate adds directly to the bicarbonate alkalinity without reacting with carbon dioxide; however, precautions must be taken not to add this chemical to a level of sodium toxicity (>3500 mg/l) (WPCF, 1987). Certain volatile fatty acids, including propionic and higher molecular weight acids, are associated with the onset of digester failure (Hill and Holmberg, 1988; Ahring, 1995). Useful parameters based on this principle are the ratio of these acids to acetic acid (Hill et al., 1987) and concentrations of iso-volatile acids (Hill and Holmberg, 1988).

3.6.3 Oxidation-Reduction Potential

Oxidation reduction potential (ORP) is a measure of the electron activity in aqueous environments and has been used on a limited basis as a performance parameter for anaerobic digestion (Dirasian et al., 1963; Gupta et al., 1994). Methanogenic bacteria are the most sensitive organisms in the biomethanogenic fermentation to elevated ORP levels, and an increase in this parameter indicates inhibition. The criticism that this parameter is sensitive to exposure of samples to oxygen during sampling and measurement and electrode fouling in anaerobic digesters (Zehnder and Stumm, 1988; Srivinas et al., 1988) was circumvented by use of a flow-through cell and intermittent cleaning of the electrodes (Gupta et al., 1994).

3.6.4 Reactor Design

Designs have been developed for in-vessel anaerobic digestion of MSW (see reviews by Cecchi et al., 1988; Cecchi and Mata-Alvarez, 1991; IEA, 1994). Each has its own benefits and constraints, and selection is dependent upon waste characteristics and personal preference. The designs are dependent upon factors such as reactor solids concentration, mixing strategy, temperature, and number of stages. Wet continuous digesters are mixed, fed continuously or semicontinuously, and operated at solids concentrations of <10%; however, since MSW is initially >80% solids, this design is often ruled out. The dry continous reactor involves semicontinuous feeding and operation at solids concentrations of 20 to 40%. The differences in design are based on mechanisms of feeding and mixing. All designs recycle effluent which is mixed with the feed for inoculation. The DRANCO design (Six and de Baere, 1992) does not mix after feeding. The VALORGA design (Saint-Joly, 1992) blast mixes with pressurized biogas. The KOMPOGAS design (Wellinger et al., 1993) is horizontal and mixed continuously at a low rate. These designs achieve high loading rates and minimize water requirements; however, mixing, feeding, and startup are challenges.

Dry batch reactors involve inoculation of a new batch of feed with contents of a previous run, and the reaction is allowed to go to completion. This design is simple and resembles an enhanced landfill. The disadvantages include process stability and materials handling related to batch operation. The BIOCEL process (Ten Brummeler et al., 1992) is an example of this design.

Several designs of the multi-stage wet reactor mix the waste with liquid, allow depolymerization to occur, and then convert the liquefied depolymerization and fermentation products to methane in high-rate, attached-film reactors. Examples of these plants are the BTA (Kubler and Schertler, 1994) and PAQUES (Hack and Brinkman, 1992) designs. They claim lower retention times and more complete conversion of solids. The major disadvantage is the complexity of design and operation.

The leach-bed design uses recycle of leachate between new and mature reactors to inoculate, wet, and provide nutrients for rapid startup of new cells. Organic acids produced during startup are conveyed via leachate to the mature reactor for conversion (Ghosh, 1984; Chynoweth et al., 1991, 1992; O'Keefe et al., 1993). This design operates at high solids (>35%), can be conducted in reactors or simple controlled landfill cells, and does not require mixing. A disadvantage is the lack of a mechanism for continuous feed. This design was developed as the SEBAC process at the University of Florida and is ready for demonstration on a commercial scale.

There are three major advantages to multi-phase designs. The first involves improved stability. In a single combined phase digester, overloading and inhibitors result in accumulation of volatile organic acids for which populations of organisms are not available to metabolize. Enrichment for these organisms can take months. In a two-phase system, formation of acids is encouraged in the acid phase; therefore, the methane phase is constantly receiving acids to encourage maintenance of high populations of these organisms. In other words, the acid phase is an intentionally imbalanced digester which is resistant to further imbalances resulting from overloading or inhibitors. The second advantage is that the slow-growing populations of microorganisms (acid users and methanogens) can be concentrated in biofilms, thus permitting short retention times for this reactor. This reduces the overall reactor volume requirement, including both stages. The third advantage is that most of the methane is produced in the methane-phase digester and the methane content of this gas is higher because of the release of much of the carbon dioxide in the acid phase. This advantage facilitates gas use by localizing its production and increasing its methane content.

3.7 COMMERCIALIZATION AND PRODUCT USE

The commercial application of anaerobic digestion to MSW is just beginning to emerge. A recent report (IEA, 1994) from the International Energy Agency has indicated that 20 demonstration plants are in operation and another 26 plants are in the planning stage or are under construction. These plants range in size from 500 to 100,000 tons per year and represent a variety of designs and sizes. The numbers of existing plants by country are Germany (10), India (3), Denmark (2), Netherlands (3), Switzerland (3), Belgium (2), Sweden (3), France (1), Italy (1), Austria (1), and Tahiti (1).

Several barriers to commercialization of anaerobic digestion of MSW still exist. The major barrier is that tipping fees (cost per ton to treat waste, usually excluding source or mechanical separation) are still influenced by low costs of landfilling. Tipping fees in the range of $50 to $70 per ton are estimated for the anaerobic digestion option. However, as liabilities and lack of public acceptance of landfills increase, anaerobic digestion may become more attractive. Another barrier is that anaerobic digestion of MSW is incomplete, resulting in a residue that must then be disposed. If free of glass, plastics, and other undesired components, the residue may be used as a soil amendment; however, the only way of achieving this "clean" compost is through costly pre- and post-treatment separation operations or source separation.

Biogas is generated from the OFMSW at a typical yield of 0.4 m^3/kg VS added. The composition is typically 55% methane and 45% carbon dioxide, with traces of hydrogen, hydrogen sulfide, and water vapor. This gas is combustible without purification and can be used directly for heating, cooking, and running generators and internal combustion engines. These uses often require some passage through a condensation trap to reduce the water content. Biogas also can be upgraded by removal of carbon dioxide and hydrogen sulfide and compressed for use in motor vehicles or distribution into the gas pipline (Constant et al., 1989; Walsh et al., 1988). A typical plant treating the OFMSW from a population of 100,000 in the U.S. could be expected to generate about 50,000 cubic meters of methane per day (Chynoweth, 1994).

Anaerobic digestion typically results in a 50% reduction in VS. The extent of conversion is dependent upon the feedstock and is similar to that obtained by aerobic composting operated at comparable residence times. Both processes require "curing" for a couple of weeks to oxidize reduced compounds such as ammonia, sulfide, and other reduced inorganic and organic compounds. The quality of the compost is also related to content of glass, plastics, and heavy metals present which is dictated by pre- and post-separation operations. Experience has shown that the cleanest compost is obtained by source separation of the organic fraction prior to digestion (IEA, 1994). A typical application for compost is about 6.8 tons per hectare per year which would require about 1800 ha for sustained application of compost resulting from anaerobic digestion of the OFMSW generated by a population of 100,000 (Chynoweth, 1994). This compost would function primarily as a soil conditioner and not as a major source of nutrients (Kayhanian and Tchobanoglous, 1993; Rivard et al., 1995b).

Destruction of human, animal, and plant pathogens during treatment of organic wastes is a major concern for subsequent use of the compost. In general, most studies have shown that anaerobic digestion results in reduction in numbers of pathogenic organisms (Berg and Berman, 1980; Engeli et al., 1993; Stukenberg et al., 1994; Bendixen, 1994). Destruction of these organisms is related to temperature and is only effective at thermophilic temperatures. Berg and Berman (1980) reported that coliform organisms were good indicators of survival of enteric viruses and that fecal streptococci would be better indicators. Bendixen (1994) showed that most pathogens were killed in thermophilic solid waste digesters, including bacteria, viruses, and parasites.

Engeli et al. (1993) showed that plant pathogens and weed seeds not killed by aerobic composting were significantly reduced by thermophilic digestion. Mesophilic temperatures did not result in effective reduction of pathogens. These results point to the advantages of anaerobic digestion and thermophilic temperatures for effective pathogen reduction.

3.8 CONCLUSION

A number of environmental benefits are provided by anaerobic digestion of MSW. The major benefit is the conversion of about 50% (weight and volume) of the organic fraction stream into useful fuel and a soil amendment. The process itself is environmentally benign and does not generate toxic products. MSW can be treated as feed blends with other wastes from agricultural, domestic sludge, and industrial sources. The process provides for the controlled methane fermentation of wastes, thus preventing uncontrolled release of methane into the atmosphere (see Chapter 1). The energy requirements for anaerobic digestion and associated operations are typically less than 10% of the methane product. The ideal anaerobic digestion feed is the separated organic fraction of MSW, the preparation of which would encourage and facilitate recycling of other fractions, including glass, metals, plastics, and paper. Finally, thermophilic anaerobic digestion results in reduction of human and plant pathogens.

Knowledge of the microbiology of anaerobic digestion is lacking, particularly as it applies to the decomposition of MSW. Information on this process is limited primarily to factors that affect overall methanogenesis and activities and interactions of physiological groups. The best understood groups are the methanogens and syntrophic organisms involved in interspecies hydrogen and acetate transfer. Most of this information has been derived from studies of a variety of ecosystems, and the same mechanisms are inferred for MSW. The major organic constituents of MSW are cellulose, hemicellulose, and lignin. These substrates are bound in particles, and their solubilization and depolymerization is the rate-limiting step of MSW decomposition. In spite of this, organisms involved in this important step are the least studied and understood. With the advent and ease of modern molecular techniques, it should be possible to isolate and characterize the organisms involved in the first steps of the fermentation. This should lead to a better understanding of this step and to methods for enhancing the extent and rate of conversion.

The importance of anaerobic organisms in the degradation of halogenated organic and other xenobiotic compounds should be reemphasized. These compounds may occur in the mix of MSW, and their degradation is relevant to prevention of future environmental impact. The optimization of anaerobic digestion of MSW and other substrates is dependent upon understanding the microbial mechanisms involved and the application of this knowledge for improved design, operation, evaluation of performance, and process control.

REFERENCES

Adney, W.S., C.J. Rivard, K.Grohmann, and M.E. Himmel. 1989. Detection of extra-cellular hydrolytic enzymes in the anaerobic digestion of municipal solid waste. *Biotech. Appl. Biochem.* 11:387–400.

Ahring, B.K. 1995. Methanogenesis in thermophilic biogas reactors. *Antonie van Leeuwenhoek* 67:91–102.

Al-Yousfi, A.B., and F.G. Pohland. 1993. Modeling of leachate and gas generation during accelerated biodegradation at controlled landfills, in *Proceedings of 31st Annual International Solid Waste Exposition,* Solid Waste Association of North America, San Jose, CA, pp. 275–290.

Anderson, G.K., B. Kasapgil, and O. Ince. 1994. Microbiological study of two-stage anaerobic digestion during start-up. *Water Res.* 28:2383–2392.

Andrews, J.F., and S.P. Graef. 1971. Dynamic modeling and simulation of the anaerobic digestion process, in *Anaerobic Biological Treatment Processes, Advances in Chemistry.* Series 105. American Chemical Society, Washington D.C., pp. 126–162.

Angelidaki, I., and B.K. Ahring. 1994. Anaerobic thermophilic digestion of manure at different ammonia loads: effect of temperature. *Water Res.* 28:727–731.

Armiger, W.B., J.F. Forro, L.M. Montalve, J.F. Lee, and D.W. Zabriskie. 1986. The interpretation of on-line measurements of intra-cellular NADH in fermentation processes. *Chem. Eng. Commun.* 45:197–206.

Balch, W.E., G.E. Fox., L.J.Magrum, C.R. Woese, and R.S. Wolfe. 1979. Methanogens: reevaluation of a unique biological group. *Microbiol. Rev.* 43:260–296.

Baldwin, D.J. 1993. *Appraisal of Farm Waste Options.* Energy Technology Support Unit (ETSU), Harwell Laboratory, Oxfordshire, U.K.

Barbee, G.C. 1994. Fate of chlorinated aliphatic hydrocarbons in the vadose zone and ground water. *Ground Water Monitor. Rem.* 14:129–140.

Barlaz, M.A., R.K. Ham, and D.M. Schaefer. 1990. Methane production from municipal refuse: a review of enhancement techniques and microbial dynamics. *CRC Crit. Rev. Envir. Control* 19:557–584.

Beaty, P.S., and M.J. McInerney. 1989. Effects of organic acid anions on the growth and metabolism of *Syntrophomonas wolfei* in pure culture and in defined consortia. *Appl. Environ. Microbiol.* 55:977–983.

Béguin, P. and J.P. Aubert. 1994. The biological degradation of cellulose. *FEMS Microbiol. Rev.* 13:25–58.

Bendixen, H.J. 1994. Safeguards against pathogens in Danish biogas plants, in T.J. Britz and F.G. Pohland (Eds.), *Anaerobic Digestion VII*, Vol. 30 (12). Elsevier Science, Amsterdam, pp. 171–180.

Berg, G. and D. Berman. 1980. Destruction by anaerobic mesophilic and thermophilic digestion of viruses and indicator organisms indigenous to domestic sludges. *Appl. Envir. Microbiol.* 39:361–368.

Blakey, N.C., R. Cossu, P.F. Maris, and F.E. Mosey. 1992. Anaerobic lagoons and UASB reactors: laboratory experiments, in T.H. Christensen, R. Cossu, and R. Stegmann (Eds.), *Landfilling of Waste: Leachate.* Elsevier Science, London, pp. 245–264.

Boone, D., and R. Mah. 1987. Transitional bacteria, in D.P. Chynoweth and R. Isaacson (Eds.), *Anaerobic Digestion of Biomass.* Elsevier Science, London, pp. 35–48.

Brigmon, R.L., D.P. Chynoweth, J.C. Yang, and S.G. Zam. 1994. An enzyme-linked immunosorbent assay (ELISA) for detection of *Clostridium alderichii* in anaerobic digesters. *J. Appl. Bacteriol.* 77:448–465.

Cecchi, F., and J. Mata-Alvarez. 1991. Anaerobic digestion of municipal solid waste: an up-to-date review, in H.W. Doelle, D.A. Mitchell, and E. Rolz (Eds.), *Solid Substrate Cultivation*, Elsevier Science, Amsterdam, pp. 369–384.

Cecchi, F., J. Mata-Alvarez, and F.G. Pohland (Eds.). 1993. *Anaerobic Digestion of Solid Waste*, Vol. 27. Pergamon Press, Oxford.

Cecchi, F., P. Traverso, J. Mata-Alvarez, J. Clancy, and C. Zaror. 1988. Anaerobic digestion of municipal solid waste in Europe. *Biomass* 16:257–284.

Cecchi, F., P.G. Traverso, J. Mata-Alvarez, F. Medici, and G. Fazzini. 1990. A new approach to the kinetic study of anaerobic digestion of the organic fraction of municipal solid waste. *Biomass* 23:79–102.

Chaudhry, G.R. and S. Chapalamadugu. 1991. Biodegradation of halogenated organic compounds. *Microbiol. Rev.* 99:59–79.

Chen, T., D.P. Chynoweth, and R. Biljetina. 1990. Anaerobic digestion of municipal solid waste in a nonmixed solids-concentrating digester. *Appl. Biochem. Biotechnol.* 24/25:533–544.

Chen, Y.R., and A.G. Hashimoto. 1979. Kinetics of methane fermentation, in C.D. Scott (Ed.), *Biotechnology and Bioengineering Symposium*, No.8. John Wiley & Sons, New York, pp. 269–282.

Christensen, J. and K. Hjort-Gregersen. 1994. The commercialization of biogas production (Denmark), in *Proc. Seventh International Symp. on Anaerobic Digestion*. Capetown, South Africa.

Christensen, T.H., R. Cossu, and R. Stegmann (Eds.). 1992. *Landfilling of Waste: Leachate*. Elsevier Science, London.

Chynoweth, D.P. 1987a. Introduction, in D.P. Chynoweth and R. Isaacson (Eds.), *Anaerobic Digestion of Biomass*. Elsevier Science, New York, pp. 1–14.

Chynoweth, D.P. 1987b. Biomass conversion options, in K.R. Reddy and W.H. Smith (Eds.), *Aquatic Plants for Water Treatment and Resources Recovery*. Magnolia Publishing, Orlando, FL, pp. 621–642.

Chynoweth, D.P. 1992. Global significance of biomethanogenesis, in D. Dunnette and R. O'Brian (Eds.), *Global Environmental Chemistry*, ACS Symposium Series 483. American Chemical Society, Washington, D.C., pp. 338–351.

Chynoweth, D.P. 1994. The anaerobic option for biosolids management (abstract). American Society of Agricultural Engineers International Meeting, Atlanta, GA.

Chynoweth, D.P. 1996. Environmental impact of biomethanogenesis. *Envir. Monitor. Assess.* (in press).

Chynoweth, D.P., G. Bosch, J.F.K. Earle, R. Legrand, and K. Liu. 1991. A novel process for anaerobic composting of municipal solid waste. *Appl. Biochem. Biotechnol.* 28/29:421–432.

Chynoweth, D.P., G. Bosch, J.F.K. Earle, J. Owens, and R. Legrand. 1992. Sequential batch anaerobic composting of the organic fraction of municipal solid waste. *Water Sci. Tech.* 24:327–339.

Chynoweth, D.P., K.F. Fannin, and V.J. Srivastava. 1987. Biological gasification of marine algae, in K. Bird and P. Benson (Eds.), *Seaweed Cultivation for Renewable Resources*. Elsevier Science, New York, pp. 285–303.

Chynoweth, D.P., and R. Isaacson (Eds.) 1987. *Anaerobic Digestion of Biomass*. Elsevier Science, London.

Microbiology of Solid Waste

Chynoweth, D.P. and R.A. Mah. 1971. Volatile acids formation in sludge digestion. *Am. Chem. Soc. Adv. Chem. Ser.* 105:41–54.

Chynoweth, D.C., S.A. Svoronos, G. Lyberatos, J.L. Harmon, P. Pullammanappallil, J.M. Owens, and M.J. Peck. 1994. Real-time expert system control of anaerobic digestion, in T.J. Britz and F.G. Pohland (Eds.), *Anaerobic Digestion VII*, Vol. 30(12). Pergamon Press, Oxford, pp. 21–30.

Chynoweth, D.P., C.E. Turick, J.M. Owens, D.E. Jerger, and M.W. Peck. 1993. Biochemical methane potential of biomass and waste feedstocks. *Biomass Bioenergy* 5:95–111.

Colberg, P.J. 1988. Anaerobic microbial degradation of cellulose, lignin, oligolignols and monoaromatic lignin derivatives, in A.J.B. Zehnder (Ed.), *Biology of Anaerobic Microorganisms*, John Wiley & Sons, New York, pp. 333–372.

Collins, L.J., and A.R. Paskins. 1987. Measurement of trace concentrations of hydrogen in biogas from anaerobic digesters using an exhaled hydrogen monitor. *Water Res.* 21:1567–1572.

Constant, M., H. Naveau, J.L. Ferrere, and E.J. Nynes. 1989. *Biogas: End Use in the European Community.* Elsevier Science, London.

Contois, D.E. 1959. Kinetics of bacterial growth. Relationship between population density and specific growth rate of continuous cultures. *J. Gen. Microbiol.* 21:40–50.

Costello, D.J., P.F. Greenfield, and P.L. Lee. 1991. Dynamic modeling of a single-stage high-rate anaerobic reactor, I. Model derivation. *Water Res.* 25:847–858.

deRosa, M., A. Gambacorta, and A. Gliozzi. 1986. Structure, biosynthesis, and physicochemical properties of archaebacterial lipids. *Microbiol. Rev.* 50:70–80.

Dirasian, H.A., A.H. Molof, and J.A. Borchardt. 1963. Electrode potentials developed during sludge digestion. *J. Water Poll. Control Fed.* 35:424–439.

Eastman, J. A. and J.F. Ferguson. 1981. Solubilization of particulate organic carbon during the acid phase of anaerobic digestion. *J. Water Poll. Control Fed.* 53:352–366.

Edelmann, W., and H. Engeli. 1993. Combined digestion and composting of organic industrial and municipal wastes in Switzerland. *Water Sci. Tech.* 27:160–182.

Elder, D.J.E. and D.J. Kelly. 1994. The bacterial degradation of benzoic acid and benzenoid compounds under anaerobic conditions: unifying trends and new perspectives. *FEMS Microbiol. Rev.* 13:441–468.

Engeli, H., W. Edelmann, J. Fuchs, and K. Rottermann. 1993. Survival of plant pathogens and weed seeds during anaerobic digestion. *Water. Sci. Tech.* 27:69–76.

Eriksson, K-E, R.A. Blanchette, and P. Ander. 1990. *Microbial and Enzymatic Degradation of Wood and Wood Components.* Springer-Verlag, Berlin.

Fannin, K.F. 1987. Start-up, operation, stability, and control, in D.P. Chynoweth and R. Isaacson (Eds.), *Anaerobic Digestion of Biomass*, Elsevier Science, London, pp. 171–196.

Fannin, K.F., and R. Biljetina. 1987. Reactor designs, in D.P. Chynoweth and R. Isaacson (Eds.), *Anaerobic Digestion of Biomass.* Elsevier Science, London, pp. 141–169.

Gander, J.E., S.J. Bonetti, J.R. Brouillette, and C.A. Abbas. 1993. Depolymerization of structural polymers: use of phosphodiesterases and glycohydrolases. *Biomass Bioenergy* 5:223–239.

Ghosh, S. 1984. Solid-phase methane fermentation of solid wastes, in *Proc. Eleventh American Society of Mechanical Engineers National Waste Processing Conference*, Orlando, FL.

Gilkes, N.R., D.G. Kilburn, R.C. Miller, Jr., and R.A.J. Warren. 1991. Bacterial cellulases. *Bioresource Technol.* 36:21–35.

Grau, P., M. Dohanyas, and J. Chudoba. 1975. Kinetics of multicomponent substrate
removal by activated sludge. *Water Res.* 9:637–642.

Gujer, W., and A.J.B. Zehnder. 1983. Conversion processes in anaerobic digestion.
Water Sci. Technol. 15:127–167.

Gupta, M., A. Gupta, M.T. Suidan, G.D. Sayles, and J.R.V. Flora. 1994. ORP measurement in anaerobic systems using flow-through cell. *J. Envir. Eng.*
120:1640–1645.

Hack, P.J.F.M., and J. Brinkman. 1992. A new process for high performance digestion,
in F. Cecchi, J. Mata-Alverez, and F.G. Pohland (Eds.), Proc. International Symp.
Anaerobic Digestion of Solid Waste. Venice, Italy.

Harmon, J.L., S.A. Svoronos, G. Lyberatos, and D.P. Chynoweth. 1993. Adaptive temperature optimization of continuous anaerobic digesters. *Biomass Bioenergy* 4:1–7.

Haskoning. 1994. *Conversion Techniques for VHF-Biowast, Developments in 1992.*
Haskoning, Barbarossastraat 35. Nijmegen, The Netherlands.

Hendrick, D.B., and D.C. White. 1993. Application of analytical microbial ecology to
the anaerobic conversion of biomass to methane. *Biomass Bioenergy* 5:247–259.

Hendrick, D.B., A. Vass, B. Richards, W. Jewell, J.B. Guckert, and D.C. White. 1991a.
Starvation and overfeeding stress on microbial activities in high-solids high-yield
methanogenic digesters. *Biomass Bioenergy* 1:75–82.

Hendrick, D.B., T. White, J.B. Guckert, and D.C. White. 1991b. The effects of oxygen
and chloroform on microbial activities in a high-solids, high-productivity anaerobic biomass reactor. *Biomass Bioenergy* 1:207–212.

Hickey, R.F., J. Vanderwielen, and M.S. Switzenbaum. 1987. Production of trace levels
of carbon monoxide during methanogenesis on acetate and methanol. *Biotechnol.
Lett.* 9:63–66.

Hill, D.T., S.A. Cobb, and J.B. Bolte. 1987. Using volatile fatty acid relationships to
predict anaerobic digester failure. *ASAE Trans.* 30:496–501.

Hill, D.T., and R.D. Holmberg. 1988. Long chain fatty acid relationships in anaerobic
digestion of swine waste. *Biological Wastes* 23:195–214.

Hungate, R.E. 1967. A roll-tube method for cultivation of strict anaerobes, in J.R.
Norris and D.W. Ribbons (Eds.), *Methods in Microbiology*, Vol. 2B. Academic
Press, New York, pp. 117–132.

IEA. 1994. *Biogas from Municipal Solid Waste: Overview of Systems and Markets for
Anaerobic Digestion of MSW*. Report of International Energy Agency Task XI:
Conversion of MSW to Energy; Activity 4: Anaerobic Digestion of MSW, Ministry
of Energy, Copenhagen.

Jackson, G.W., and D.F. James. 1986. The permeability of fibrous porous media. *Can.
J. Chem. Eng.* 64:364

Jain, S., A.K. Lala, S.K. Bhatia, and A.P. Kudchadker. 1992. Modeling hydrolysis
controlled anaerobic digestion. *J. Chem. Tech. Biotechnol.* 53:337–344.

Jewell, W.J., R.J. Cummins, and B.K. Richards. 1993. Methane fermentation of energy
crops: maximum conversion kinetics and *in situ* biogas purification. *Biomass
Bioenergy* 5:261–278.

Kamagata, Y., N. Kitagawa, M. Taskai, K. Nakamura, and E. Mikami. 1992. Degradation of benzoate by an anaerobic consortium and some properties of a hydro-
genotrophic methanogen and sulfate-reducing bacterium in the consortium. *J.
Ferment. Bioeng.* 73:213–218.

Kane, M.D., J.M. Stromley, L. Raskin, and D.A. Stahl. 1991. *Molecular Analysis of the Phylogenetic Diversity and Ecology of Sulfidogenic and Methanogenic Biofilm Communities.* Abstract Q-195. Abstr. Annual Meeting American Society of Microbiology, p. 309.

Kayhanian, M. 1994. Performance of a high-solids anaerobic digestion process under various ammonia concentrations. *J. Chem. Tech. Biotechnol.* 59:349–352.

Kayhanian, M., and S. Hardy. 1994. The impact of four design parameters on the performance of a high-solids anaerobic digestion of municipal solid waste for fuel gas production. *Environ. Technol.* 15:557–567.

Kayhanian, M., and G. Tchobanoglous. 1993. Characteristics of humus produced from the anaerobic composting of the biodegradable organic fraction of municipal solid waste. *Environ. Technol.* 14:815–829.

Kelly, C.A., and D.P. Chynoweth. 1979. Methanogenesis: measurement of chemo-organotrophic (heterotrophic) activity in anaerobic lake sediments, in J.W. Costerton and R.R. Colwell (Eds.), *Native Aquatic Bacteria: Enumeration, Activity, and Ecology.* ASTM 695. American Society for Testing and Materials, Philadelphia, PA, pp. 164–179.

Ke-yun, D., Z. Yi-zhang, and W. Li-bin. 1988. The role of biogas development in improving rural energy, in E. Hall and P.N. Hobson (Eds.), *Anaerobic Digestion 1988.* Pergamon Press, Oxford, pp. 295–302.

Krzycki, J.A., W.R. Kenealy, M.J. DeNiro, and J.G. Zeikus. 1987. Stable isotope fractionation by *Methanosarcina barkeri* during methanogenesis from acetate, methanol, or carbon dioxide. *Appl. Environ. Microbiol.* 53:2597–2599.

Kubler, H., and C. Schertler. 1994. Three-phase anaerobic digestion of organic wastes, in T.J. Britz and F.G. Pohland (Eds.), *Anaerobic Digestion VII,* Vol. 30(12). Pergamon Press, Oxford, pp. 367–374.

Labib, F., J.F. Ferguson, M.M. Benjamin, M. Merigh, and N.L. Ricker. 1993. Mathematical modeling of an anaerobic butyrate degrading consortia: predicting their response to organic overloads. *Environ. Sci. Tech. Res.* 27:2673–2684.

Lagerkvist, A. and Chen, H. 1993. Control of two-step anaerobic degradation of municipal solid waste (MSW) by enzyme addition. *Water Sci. Technol.* 27:47–56.

Lamed, R. and E.A. Bayer. 1988. The cellulosome of *Clostridium thermocellum. Adv. Appl. Microbiol.* 33:1–46.

Lee, M.J., P.J. Schreurs, A.C. Messer, and S.H. Zinder. 1987. Association of methanogenic bacteria with flagellated protozoa from a termite hindgut. *Curr. Microbiol.* 15:337–371.

Lin, C-Y. 1993. Effect of heavy metals on acidogenesis in anaerobic digestion. *Water Res.* 27:147–152.

Linden, J.C. and M. Shiang. 1991. Bacterial cellulases: regulation of synthesis, in G.F. Leatham and M.E. Himmel (Eds.), *Enzymes in Biomass Conversion.* ACS Symposium Series 460. American Chemical Society, Washington, D.C., pp. 331–348.

Ljungdahl, L.G. and K-E. Eriksson. 1985. Ecology of microbial cellulose degradation, in K.C. Marshall (Ed.), *Advanced Microbiology and Ecology,* Vol. 8. Plenum Press, New York, pp. 237–299.

Ljungdahl, L., B. Pettersson, K-E. Eriksson, and J. Wiegel. 1983. A yellow affinity substance involved in the cellulolytic system of *Clostridium thermocellum. Curr. Microbiol.* 9:195–200.

Lusk, P. 1994. *Methane Recovery From Animal Manures: A Current Opportunities Casebook,* Vols. I and II. Prepared for National Renewable Energy Laboratory, Golden, CO.

Macario, A.J.L,. and E.C. de Macario. 1983. Antigenic fingerprinting of methanogenic bacteria with polyclonal antibody probes. *Syst. Appl. Microbiol.* 4:451–458.

Macario, A.J.L., and E.C. de Macario. 1988. Quantitative immunological analysis of the methanogenic flora of digesters reveals a considerable diversity. *Appl. Environ. Microbiol.* 54:79–86.

Macario, A.J.L., and E.C. de Macario. 1993. Immunology of methanogenic bacteria. *Biomass Bioenergy* 5:203–213.

Macario, A.J.L., J.K. Earle, D.P. Chynoweth, and E.C. de Macario. 1989. Distinctive patterns of methanogenic flora determined with antibody probes in anaerobic digesters of different characteristics operated under controlled conditions. *Syst. Appl. Microbiol.* 12:216–222.

Macario, A.J.L., M.W. Peck, E. Conway de Macario, and D.P. Chynoweth. 1991. Unusual methanogenic flora of a wood-fermenting anaerobic bioreactor. *Appl. Bacteriol.* 71:31–37.

Mata-Alvarez, J., F. Cecchi, and P. Pavan. 1992. Substrate utilization kinetic models in the semi-dry thermophilic anaerobic digestion of municipal solid waste. *J. Envir. Sci. Health* 27:1967–1986.

McCarty, P.L. 1964. Anaerobic treatment fundamentals. *Public Works,* Sept., Oct., Nov., Dec. issues.

McCarty, P.L. 1971. Energetics and kinetics of anaerobic treatment, in *Advances in Chemistry,* Series 105. American Chemical Society, Washington, D.C., pp. 91–107.

McCarty, P.L. 1992. One hundred years of anaerobic treatment, in D.E. Hughes et al. (Eds.), *Anaerobic Digestion 1991.* Elsevier, Amsterdam, pp. 3–22.

McInerney, M. J. 1988. Anaerobic hydrolysis and fermentation of fats and proteins, in A.J.B. Zehnder (Ed.), *Biology of Anaerobic Microorganisms,* John Wiley & Sons, New York, pp. 373–416.

Monod, J. 1949. The growth of bacterial cultures. *Ann. Rev. Microbiol.* 3:371–376.

Mosey, F.E., and X.A. Fernandes. 1989. Patterns of hydrogen in biogas from the anaerobic digestion of milk-sugars. *Water Sci. Technol.* 21:87–196.

Mosey, F.E., and D.A. Hughes. 1975. The toxicity of heavy metal ions to anaerobic digestion. *J. Inst. Water Poll. Control.* 1:3–24.

Mueller, R.F., and A. Steiner. 1992. Inhibition of anaerobic digestion caused by heavy metals. *Water Sci. Technol.* 26:834–846.

Negri, E.D., J. Mata-Alvarez, C. Sans, and F. Cecchi. 1993. A mathematical model of volatile fatty acids (VFA) production in a plug-flow reactor treating the organic fraction of the municipal solid waste (MSW). *Water Sci. Technol.* 27:201–208.

Ng, H. and R.H. Vaughn. 1963. *Clostridium rubrum* sp. n. and other pectinolytic clostridia from soil. *J. Bacteriol.* 85:1104–1113.

Noike, T., G. Endo, J-E. Chang, J-I. Yaguchi, and J-I. Matsumoto. 1985. Characteristics of carbohydrate degradation and rate-limiting step in anaerobic digestion. *Biotech. Bioeng.* 27:1482–1489.

Odier, E. and I. Artaud. 1992. Degradation of lignin, in G. Winkelmann (Ed.), *Microbial Degradation of Natural Products.* Verlag Chemie, Weinheim, Germany, pp. 161–192.

O'Keefe, D.M., D.P. Chynoweth, A.W. Barkdoll, R.A. Nordstedt, J.M. Owens, and J. Sifontes. 1993. Sequential batch anaerobic composting of municipal solid waste and yard waste. *Water Sci. Technol.* 27:77–86.

Oremland, R.S., II. 1988. Biogeochemistry of methanogenic bacteria, in A.J.B. Zehnder (Ed.), *Biology of Anaerobic Microorganisms*, John Wiley & Sons, New York, pp. 641–706.

O'Rourke, J.T. 1968. Kinetics of Anaerobic Treatment at Reduced Temperatures. Ph.D. dissertation, Stanford University, Stanford, CA.

Owen, W.F., D.C. Stuckey, J.B. Healy, Jr., L.Y. Young, and P.L. McCarty. 1979. Bioassay for monitoring biochemical methane potential and anaerobic toxicity. *Water Res.* 13:485–492.

Owens, J.M., and D.P. Chynoweth. 1993. Biochemical methane potential of MSW components. *Water Sci. Technol.* 27:1–14.

Parkin, G.F., R.E. Speece, C.H.J. Yang. W.M. Kocher. 1983. Response of methane fermentation systems to industrial toxicants. *J. Water Poll. Control Fed.* 55:44–53.

Pavlostathis, S.G., and F. Giraldo-Gomez. 1991. Kinetics of anaerobic treatment: a critical review. *CRC Crit. Rev. Environ. Control* 21:411–490.

Peck, M.W., and D.B. Archer. 1989. Methods for the quantification of methanogenic bacteria. *Intl. Ind. Biotechnol.* 9:5–12.

Peck, M.W., and D.P. Chynoweth. 1990. On-line monitoring of the methanogenic fermentation by measurement of culture fluorescence. *Biotechnol. Lett.* 12:17–22.

Peck, M.W., and D.P. Chynoweth. 1992. On-line monitoring of the methanogenic fermentation. *Biotech. Bioeng.* 39:151–160.

Peres, C.S., C.R. Sanchez, C. Matumoto, and W. Schimidell. 1992. Anaerobic biodegradability of the organic fraction of the organic components of municipal solid wastes (OFMSW). *Water Sci. Technol.* 25:285–294.

Pfeffer, J.T. 1974. Temperature effects on anaerobic fermentation of domestic refuse. *Biotechnol. Bioeng.* 16:771–787.

Pfeffer, J.T., and K.A. Kahn. 1976. Microbial production of methane from municipal refuse. *Biotechnol. Bioeng.* 18:1179.

Pohland, F. 1992. Anaerobic treatment: fundamental concepts, applications, and new horizons, in J.F. Malina and F.G. Pohland (Eds.), *Design of Anaerobic Processes for the Treatment of Industrial and Municipal Wastes.* Technomic, Lancaster, PA, pp. 1–40.

Pohland, F.G., and S.R. Harper. 1985. Biogas developments in North America, in *Anaerobic Digestion 1985*, China State Biogas Association, Quangzhou, China, pp. 41–82.

Preston, III, J.F., J.D. Rice, and M.C. Chow. 1993. Pectinolytic bacteria and their secreted pectate lyases: agents for the maceration and solubilization of phytomass for fuels production. *Biomass Bioenergy* 5:215–222.

Pullammanappallil, P., S. Svoronos, G. Lyberatos, J. Owens, and D. Chynoweth. 1994. Stable performance of anaerobic digestion in the presence of high concentrations of propionic acid, in Proc. Seventh International Symp. on Anaerobic Digestion: Poster Papers. Cape Town, South Africa.

Raskin, L., L.K. Poulse, D.R. Noguera, B.E. Rittmann, and D.A. Stahl. 1994a. Quantification of methanogenic groups in anaerobic biological reactors by oligonucleotide probe hybridization. *Appl. Environ. Microbiol.* 60:1241–1248.

Raskin, L., J.M. Stromley, B.E. Rittman, and D.A. Stahl. 1994b. Group-specific 16S rRNA hybridization probes to describe natural communities of methanogens. *Appl. Environ. Microbiol.* 60:1232–1240.

Rintala, J.A. and B.A. Ahring. 1994. Thermophilic anaerobic digestion of source-sorted household solid waste: the effects of enzyme additions. *Appl. Microbiol. Biotechnol.* 40:916–919.

Rivard, C.J., W.S. Adney, and M.E. Himmel. 1991. Enzymes for anaerobic municipal solid waste disposal, in G.F. Leatham and M.E. Himmel (Eds.), *Enzymes in Biomass Conversion*, ACS Symposium Series 460. American Chemical Society, Washington, D.C., pp. 22–35.

Rivard, C.J., B.D. Kay, D.H. Kerbaugh, N.J. Nagle, and M.E. Himmel. 1995a. Horsepower requirements for high-solids anaerobic digestion. *Appl. Biochem. Biotech.* 51/52:155–162.

Rivard, C.J., R.A. Nieves, N.J. Nagle, and M.E. Himmel. 1994. Evaluation of discrete cellulase enzyme activities from anaerobic digester sludge fed a municipal solid waste feedstock. *Appl. Biochem. Biotech.* 45/46:453–460.

Rivard, C.J., J.B. Rodriguez, N.J. Nagle, J.R. Self, B.D. Kay, P.N. Soltanpour, and R.A. Nieves. 1995b. Anaerobic digestion of municipal solid waste: utility of process residues as a soil amendment. *Appl. Biochem. Biotech.* 51/52:125–135.

Robinson, R.W., and G.W. Erdos. 1985. Immuno-electron microscopic identification of *Methanosarcina* spp. in anaerobic digester fluid. *Can. J. Microbiol.* 31:839–844.

Saint-Joly, C. 1992. Three years of performance control and process monitoring in an industrial plant for MSW treatment by anaerobic digestion, in Proc. International Symp. on Anaerobic Digestion of Solid Waste, April 1992, Venice, Italy.

Senior, E., and M.T.M. Balba. 1990. Refuse decomposition, in E. Senior (Ed.), *Microbiology of Landfill Sites*. CRC Press, Boca Raton, FL.

Six, W., and L. De Baere. 1992. Dry anaerobic conversion of municipal solid waste by means of the DRANCO Process. *Water Sci. Technol.* 25:295–300.

Smith, P.H., F.M. Bordeaux, M. Goto, A. Shiralipour, A. Wilkie, J.F. Andrews, S. Ide, and M.W. Barnett. 1988. Biological production of methane from biomass, in W.H. Smith and J.R. Frank (Eds.), *Methane from Biomass: A Systems Approach*. Elsevier Science, London, pp. 291–334.

Sorensen, A.H. and B.K. Ahring. 1993. Measurements of the specific methanogenic activity of anaerobic digester biomass. *Appl. Microbiol. Biotechnol.* 40:427–431.

Speece, R.F. 1987a. Nutrient requirements, in D.P. Chynoweth and R. Isaacson (Eds.), *Anaerobic Digestion of Biomass*. Elsevier Science, London, pp. 109–128.

Speece, R.F. 1987b. Toxicity, in D.P. Chynoweth and R. Isaacson (Eds.), *Anaerobic Digestion of Biomass*. Elsevier Science, London, pp. 129–140.

Spencer, D.F. 1991. A preliminary assessment of carbon dioxide mitigation options. *Rev. Energy Environ.* 16:259–273.

Srivastava, V.J. 1987. Net energy, in D.P. Chynoweth and R. Isaacson (Eds). *Anaerobic Digestion of Biomass*. Elsevier Science, London, pp. 219–230.

Srivinas, S.P., G. Rao, and R. Mutharasan. 1988. Redox potential in anaerobic and microaerobic fermentations, in L. E. Erickson and D.Y.C. Fung, (Eds), *Handbook on Anaerobic Fermentations*. Marcel Dekker, New York, pp. 147–181.

Stamms, A.J.M., K.C.R. Cgrolle, C.T.J.J. Frijters, and J.B. van Lier. 1992. Enrichment of thermophilic propionate-oxidizing bacteria in syntrophy with *Methanobacterium thermoautotrophicum* or *Methanobacterium thermoformicicum*. *Appl. Environ. Microbiol.* 58:346–352.

Stevens, C.M., and A. Engelkemeir. 1988. Stable carbon isotope composition of methane from some natural and anthropogenic sources. *J. Geophys. Res.* 93:725–733.

Stukenberg, J.R., G. Shimp, Sr., J. Sandino, J.H. Clark, J.T. Crosse. 1994. Compliance outlook: meeting 40 CFR part 503, class B pathogen reduction criteria with anaerobic digestion. *Water Environ. Res.* 66:255–263.

Suidan, M.T., A.T. Schroeder, R. Nath, E.R. Krishnan, and R.C. Brenner. 1993. Treatment of CERCLA (Comprehensive Environmental Response, Compensation, and Liability Act) leachates by carbon-assisted anaerobic fluidized beds. *Water Sci. Technol.* 27:273–282.

Switzenbaum, M.S. 1991. Anaerobic treatment technology for municipal and industrial waters. *Water Sci. Technol.* 24:281.

Switzenbaum, M.S., E. Giraldo-Gomez, and R. Hickey. 1990. Monitoring of the anaerobic methane fermentation process. *Enzyme Microbiol. Technol.* 12:722–730.

Ten Brummeler, E., M.J. Aarnink, and I.W. Koster. 1992. Dry anaerobic digestion of solid organic waste in a BIOCEL reactor at pilot-plant scale. *Water Sci. Technol.* 25:301–310.

Ten Brummeler, E., H.C.J.M. Horbach, and I.W. Koster. 1991. Dry anaerobic batch digestion of the organic fraction of municipal solid waste. *J. Chem. Tech. Biotechnol.* 50:191–209.

Tornebene, T.G. and T.A. Langworthy. 1979. Diphytanyl and dibiphytanyl glycerol ether lipids of methanogenic bacteria. *Science* 203:51–53.

Tsao, G.T. 1984. Bacterial hydrolysis: a review, in G.L. Ferrero, M.P. Ferranti, and H. Naveau (Eds.), *Anaerobic Digestion and Carbohydrate Hydrolysis of Waste.* Elsevier Science, London, pp. 83–99.

Tsao, G.T. 1987. Pre-posttreatment, in D.P. Chynoweth and R. Isaacson (Eds.), *Anaerobic Digestion of Biomass*, Elsevier Science, London, pp. 91–108.

Tyler, S.C. 1991. The global methane budget, in J.E. Rogers and K.W.B. Whitman (Eds.), *Microbial Production and Consumption of Greenhouse Gases: Methane, Nitrogen Oxides, and Halomethanes.* American Society of Microbiology, Washington, D.C., pp. 7–38.

USEPA. 1992. Characterization of MSW in the United States: 1992 Update. Executive Summary. EPA/530-S-92-019. U.S. Environmental Protection Agency, Washington, D.C.

USEPA. 1993a. Current and Future Methane Emissions From Natural Sources (K.B. Hogan, Ed.), Report to Congress. U.S. Environmental Protection Agency, Washington, D.C.

USEPA. 1993b. Opportunities to Reduce Anthropogenic Methane Emissions in the United States (K.B. Hogan, Ed.), Report to Congress. U.S. Environmental Protection Agency, Washington, D.C.

USEPA. 1993c. Options for Reducing Methane Internationally, Volume I: Technical Options for Reducing Methane Emissions (K.B. Hogan, Ed.), Report to Congress. U.S. Environmental Protection Agency, Washington, D.C.

USEPA. 1993d. Options for Reducing Methane Internationally, Volume II: International Opportunities for Reducing Methane Emissions (K.B. Hogan, Ed.), Report to Congress. U.S. Environmental Protection Agency, Washington, D.C.

Vinzant, T.B., W.S. Adney, K. Grohmann, and C.J. Rivard. 1995. Aerobic and anaerobic digestion of processed municipal solid waste. *Appl. Biochem. Biotechnol.* 51/52:765–771.

Vogels, G.D., W.F. Hope, and C.K. Stumm. 1980. Association of methanogenic bacteria with rumen ciliates. *Appl. Environ. Microbiol.* 40:608–612.

Walsh, J.L., C.C. Ross, M.S. Smith, S.R. Harper, and W.A. Wilkins. 1988. *Handbook on Biogas Utilization*, U.S. Dept. of Energy, Southeastern Regional Biomass Energy Program, Tennessee Valley Authority, Mussel Shoals, AL.

Wang, Y-S, C.S. Byrd, and M.A. Barlaz. 1994. Anaerobic biodegradability of cellulose and hemicellulose in excavated refuse samples using a biochemical methane potential assay. *J. Ind. Microbiol.* 13:147–153.

Weimer, P.J., and J.G. Zeikus. 1977. Fermentation of cellulose and cellobiose by *Clostridium thermocellum* in the absence and presence of *Methanobacterium thermoautotrophicum*. *Appl. Environ. Microbiol.* 33:289–297.

Wellinger, A., K. Wyder, and A.E. Metzler. 1993. KOMPOGAS — a new system for the anaerobic treatment of source separated waste. *Water Sci. Technol.* 27:153–158.

Woese, C.R. 1987. Bacterial evolution. *Microbiol. Rev.* 51:221–271.

Woese, C.R., O. Kandler, and M.L. Wheelis. 1990. Towards a natural system of organisms: proposal for the domains Archaea, Bacteria, and Eucarya. *Proc. Natl. Acad. Sci. USA* 87:4756–4579.

Wojciechowicz, M. and H. Tomerska. 1971. Pectic enzymes in some pectinolytic rumen bacteria. *Acta Microbiol. Polonica Ser. A.* 3:37–61.

Wolin, M.J., and T.L. Miller. 1982. Interspecies hydrogen transfer. 15 years later. *ASM News* 48:561–565.

WPCF. 1987. *Anaerobic Sludge Digestion. Manual of Practice No. 19*, 2nd ed., Water Pollution Control Federation, Alexandria, VA.

Wu, J.H.D., and A.L. Demain. 1988. Proteins of the *Clostridium thermocellum* cellulase complex responsible for degradation of crystalline cellulose, in J-P. Aubert, P. Béguin, and J. Millet (Eds.), *Biochemistry and Genetics of Cellulose Degradation*. FEMS Symposium 43. Academic Press, London, pp. 118–131.

Wujcik, W.J., and W.J. Jewell. 1980. Dry anaerobic fermentation. *Biotech. Bioeng. Symp.* 10:43–65.

Yang, J.C., D.P. Chynoweth, D.P. Williams, and A. Li. 1990. *Clostridium aldrichii* sp. nov., a cellulolytic mesophilic inhabiting wood-fermenting anaerobic digester. *Intl. J. Syst. Bacteriol.* 40:268–272.

Zehnder, A.J.B., and W. Stumm. 1988. Geochemistry and biogeochemistry of anaerobic habitats, in A.J.B. Zehnder, *Biology of Anaerobic Microorganisms*. John Wiley & Sons, New York, pp. 1–38.

Zhang, T.C., and T. Noike. 1994. Influence of retention time on reactor performance and bacterial trophic populations in anaerobic digestion processes. *Water Res.* 28:27–36.

Zinder, S.H. 1993. Physiological ecology of methanogens, in J.G. Ferry (Ed.), *Methanogensesis: Ecology, Physiology, Biochemistry, and Genetics*. Chapman and Hall, New York, pp. 128–206.

Zinder, S.H. and M. Koch. 1984. Non-aceticlastic methanogenesis from acetate: acetate oxidation by a thermophilic syntrophic coculture. *Arch. Microbiol.* 138:263–272.

Zitomer, D.H. and Speece, R.E. 1993. Sequential environments for enhanced biotransformation of aqueous contaminants. *Environ. Sci. Technol.* 27:227–244.

4

Composting of Municipal Solid Waste and its Components

Frederick C. Miller

CONTENTS

4.1 Introduction 116
    4.1.1 Definition of Composting 116
    4.1.2 General Sequence of Events During Composting.................................116
    4.1.3 Processing Goals 119
    4.1.4 Application of Composting to Waste Treatment 119
4.2 Microbiology of Composting 120
    4.2.1 Investigation of Composting Populations: Synecological
          and Autoecological Approaches...120
    4.2.2 Historical Development of Composting Microbiology 121
    4.2.3 Population Trends During Composting 122
          4.2.3.1 Bacteria..122
          4.2.3.2 Actinomycetes 124
          4.2.3.3 Fungi...125
    4.2.4 Development and Roles of Different Populations...................................128

4.3 Microbial Ecology of Composting 129
    4.3.1 Composting as a Rapidly Changing System 129
    4.3.2 Composting as a Matric Phase Ecosystem..130
          4.3.2.1 Structure of the Matrix 130
          4.3.2.2 Effect of the Matrix on Phenotypic Expression 131
    4.3.3 Interactions Between Microbial Activity and the Physical
          and Chemical Environment..132
          4.3.3.1 Heat Evolution 133
          4.3.3.2 Temperature 134
          4.3.3.3 Oxygen and Gas Exchange..135
          4.3.3.4 Moisture 136
          4.3.3.5 pH and Ammonia 137
          4.3.3.6 Substrate Changes 139

0-8493-8361-7/96/$0.00+$.50
© 1996 by CRC Press, Inc.

4.4. Formation of Odorous Compounds 140

4..5 Process Management..142
    4.5.1 Process Strategy 142
    4.5.2 Process Configuration 142
        4.5.2.1 Windrows...142
        4.5.2.2 Aerated Windrows...143
        4.5.2.3 Aerated Static Piles 143
        4.5.2.4 Bins, Troughs, Enclosed Reactors, and Silos............................143
        4.5.2.5 Tunnels 144

4.6. Conclusions 145

References 145

4.1 INTRODUCTION

4.1.1 Definition of Composting

Composting is a microbial decomposition process. The poor heat conductance of the physical matrix combined with the high density of available substrate causes these systems to accumulate metabolically released heat. Accumulation of heat leads to a characteristic temperature increase. Higher temperatures tend to increase rates of microbial metabolic activity until inhibitory, high temperatures are reached. In recent years, the practice of ventilating the composting masses to remove heat and thereby avoid high temperatures has become common. Composting is generally managed as an aerobic process. Unlike aqueous systems, composting systems are prone to spatial heterogeneity both in the larger scale and at the microsite level within aggregates. This tendency to spatial heterogeneity is significant in that it fosters a diversity of ecological niches for microorganisms.

Our current understanding of composting microbiology is derived from disciplines including bacteriology; mycology; plant pathology; soil science; civil, agricultural, and mechanical engineering; and microbial ecology. This historical synthesis of information is now also being applied to bioremediation, biofiltration, and solid state production of biological cultures and products.

4.1.2 General Sequence of Events During Composting

Composting progresses through a sequence of stages that have been described both in general population terms (Bagstam, 1978, 1979; Macauley et al., 1993) and operationally (Miller et al., 1980). In general, composting proceeds through a successional dominance of nonactinomycete bacteria, followed by actinomycetes, then fungi; the simplest substrates are metabolized quickly, while substrates which are more complex or difficult to metabolize remain. A typical time course of changes in microbial populations during composting is illustrated in Figure 1. Composting can be viewed as an ecosystem initially favoring extreme "r" strategists (rapid growth when resources are temporarily abundant) progressing to a habitat suitable for "K" strategists (slow growth on recalcitrant nutrients) (Atlas and Bartha, 1981). For example, in the composting of wheaten straw, the neutral sugars from cell wall polysaccharides are degraded initially, while the recalcitrant lignin is not decomposed in the earlier stages of composting (Iiyama et al., 1994).

Microbial activity during composting follows a sequence of events described by Finstein et al. (1983). The first stage of composting can be described as the temperature ascent stage, where temperatures rise from ambient into the thermophilic range. Bacterial populations, which can multiply very rapidly while utilizing simple and readily available substrates, dominate this phase. Mesophilic bacteria at first predominate, but at about 45°C mesophilic populations die off and thermophilic bacterial populations become dominant. Rates of bacterial growth and activity peak during this early period.

A second stage of steady thermophilic temperatures follows that can reflect two very different conditions. If ventilative heat removal is not used to limit temperature increase, the upper temperature limits of the thermophilic bacteria will be exceeded, and overall microbial activity will be reduced to very low levels. If ventilative heat removal is used to limit an increase in temperature, then optimal temperatures for thermophilic bacterial activity can be maintained, and overall rates of bacterial activity will remain high. Normal temperatures achieved during this second stage (above 50°C) allow thermophilic bacteria to be active, but neither fungi nor actinomycetes can withstand temperatures as high as the thermophilic bacteria.

A third stage of composting is the declining temperature stage wherein temperatures drop from around 50°C to ambient. Thermophilic bacteria have used up the most readily available substrates, and bacterial metabolic activity can no longer liberate heat fast enough to maintain high temperatures. Operationally, process management often enters into a less structured phase somewhere during this period, referred to as curing. In curing, composting materials are removed from more intensive process control to a bulk storage condition where slow microbial changes continue. Sometimes curing is carried out with ventilation or periodic turning of the compost mass to increase microbial activity, but large, unventilated stockpiles are also common. Early in this stage, actinomycete populations increase as temperatures begin to decline and the remaining, more complex substrates can be attacked by extracellular enzymes. As temperatures drop below 35°C and the remaining substrates are ever more resistant to decomposition, fungal populations with the enzymatic ability to degrade the most difficult substrates become dominant. Overall microbial activity drops progressively to very low rates, because the remaining substrate is resistant to decomposition.

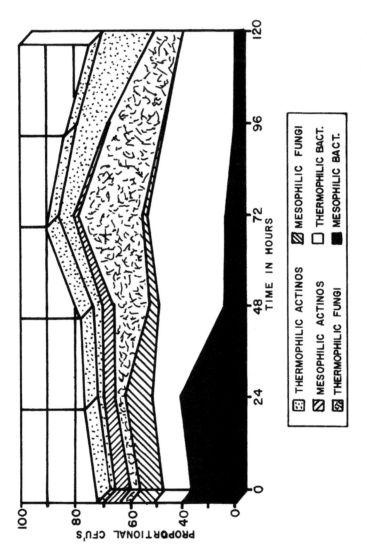

FIGURE 1 Relative changes observed in characteristic microbial populations during composting. (Adapted from Joshua R.S., B.J. Macauley, and C.R. Hudson. 1994. *BioCycle* 35:46-47. With permission.)

4.1.3 Processing Goals

Processing goals of solid waste composting include the reduction of putrescible materials, pathogen reduction, reduction of weight and volume, reduction of existing or potential odors, and a reduction in moisture content. Processing goals will vary significantly dependent upon end use. During processing, problems may occur with odors, contaminated runoff, animal vectors, noise, and release of dusts and spores. Poorly prepared composts can actually promote the regrowth of potential pathogens (Burge et al., 1987).

In a number of applications, the maturity of the product is very important. Maturity is a complex issue, however, as there are many different aspects to maturity that occur at different rates dependent on processing (Miller, 1994). Maturity is characterized by a number of parameters including (1) a reduction of putresciblity for storage or general disposal, (2) a structural change such as an increase in bulk density for landfill cover, (3) an increase in available nutrients for horticultural application, (4) reduction of pathogens for public health purposes, and other factors. Maturity is defined operationally and does not have a specific fundamental definition.

4.1.4 Application of Composting to Waste Treatment

Composting has become a useful treatment process for almost every kind of biodegradable waste. Municipal solid waste (MSW) is composted by itself, or in combination with biosolids (sewage sludge), yard waste, agricultural residues, and food waste. The nature of what is composted as MSW has changed significantly because of improved separation technologies and better definition of the "compostable" fraction.

In the U.S. in 1988 there were eight operating MSW composting facilities, 75 MSW projects under development, and perhaps 150 sewage sludge composting facilities operational or planned (Goldstein et al., 1988; Goldstein, 1989). At the end of 1993, there were 321 biosolids (sewage sludge) composting facilities planned or in operation (Goldstein and Steuteville, 1993). Yard waste composting facilities in the U.S. expanded from about 800 in 1988 to 3014 by the end of 1993 (Steuteville, 1994). In Europe in 1988 there were at least 23 composting plants for refuse, most combining refuse and sewage sludge (Bardos and Lopez-Real, 1988). Additional facilities have been built in Europe since 1988, especially for agricultural and food waste.

Early interest in solid waste composting developed in the 1950s; however, implementation of composting for solid waste was inhibited by low tipping fees at landfills, cheap incineration, and a lack of scientific and engineering knowledge to design and run composting operations well. Understanding the importance of separation to reduce heavy metal contamination of compost (Richard and Woodbury, 1992) and more recent developments in physical separation and process management (Richard, 1992) have removed many of the compost contamination issues that plagued earlier implementation.

Processing problems include odors, long processing times, products unsuitable for the desired end use, inadequate moisture control, and poor materials handling. Many composting problems can be mitigated by changes in design and management (Finstein et al., 1987a,b,c,d). The most significant management change in recent years has been the widespread adoption of temperature feedback controlled ventilation. Improvements in process management have broadened the potential application of composting as a MSW treatment method.

4.2 MICROBIOLOGY OF COMPOSTING

4.2.1 Investigation of Composting Populations: Synecological and Autoecological Approaches

Synecological and autoecological approaches are useful in the investigation of composting ecosystems. Synecological approaches focus on broad population changes and general interactions between populations and the environment in which activity is occurring. They include studies of thermodynamic and kinetic changes during composting, and the effect of temperature, moisture content, or other factors on rates of microbial activity. Autoecological investigations focus on ecological investigations of genus or species specific populations within the environment. Autoecological approaches have been used to follow the fate of pathogens during composting, the development of desired populations in applications such as mushroom composting, or the production of disease suppressive composts. In practical application, synecological approaches have been of greater relevance in composting investigations because such information is more directly useful in process management. Furthermore, because of the dynamic nature of composting ecosystems, autoecological investigations alone are difficult to interpret.

Investigations into populations in composting ecosystems have had different goals. Most studies have been restricted to population subsets of interest to individual investigators. Mycologists have tended to look for fungi, while most microbiologists have specifically looked for actinomycetes and other bacteria. Some investigators have been specifically interested in the populations common to the early high temperature stages of the composting process, while others have focused on populations common to curing materials.

Earlier work focused on the identification of specific organisms recovered from composting samples (Waksman et al., 1939a,b). More recent work has tended to focus on the activity of broad classes of organisms (Bagstam, 1979; Joshua et al., 1994). Better understanding of the autoecology of specific organisms and the physical ecology of composting ecosystems allows for a more complete interpretation of the role of microbial populations during the various stages of the composting process.

Analyzing populations in composting materials is challenging. Enumeration of fungal biomass is still difficult, making it hard to determine the specific activity of fungi during composting. There are many spore-forming bacteria that can be found in composting materials. Little is known of the role of actinomycetes in modern, environmentally controlled composting systems. There has been a particular lack of work in the isolation of facultatively anaerobic thermophilic organisms that might have functional roles in anaerobic niches.

4.2.2 Historical Development of Composting Microbiology

Composting microbiology developed from three somewhat different traditions that were later forged into the basis of our current knowledge. One line of investigation was concerned with biological self-heating (referred to as *thermogenesis* in the older literature). Browne (1933) concluded that the self-heating occurring in hay piles was of biological origin, as distinguished from chemical self-heating. Norman and colleagues (Norman, 1930; Carlyle and Norman, 1941; Norman et al., 1941) determined the time course and rates of self-heating of straws and other agricultural materials using an adiabatic apparatus. Norman (1930) had specific interests in the role of fungi in initiating self-heating in plant residues (lignocelluloses). In contained plant residues with low nitrogen, fungi were the dominant initiators of self-heating. In a series of papers, Rothbaum and Dye investigated self-heating in wool, a protein (Rothbaum, 1961, 1963; Dye and Rothbaum, 1964; Rothbaum and Dye, 1964; Dye, 1964). Rothbaum measured heat evolution and microbial activity in the thermophilic range. Working within the thermophilic range, activity was dominated by species of *Bacillus* at temperatures exceeding 55°C.

Walker and Harrison (1960), working also with wool, investigated the role of temperature in heat evolution. They found that heat evolution increased with temperature for mesophilic populations, until thermal death occurred, and that this trend was then repeated for the thermophilic populations. They also demonstrated a drop in heat evolution rates at about 45°C, indicating that mesophilic and thermophilic populations do not overlap in a smooth continuum. Walker and Harrison (1960) found that optimal mesophilic activity occurred at about 37°C, and optimal thermophilic activity occurred at about 55°C.

A second line of investigation was the identification of specific microbes and their role in composting. Early studies were initiated by workers interested in mushroom composting and decomposition of agricultural materials, while work on MSW composting came later. This second line of investigation was best characterized by the work of Waksman and colleagues (Waksman and Hutchings, 1937; Waksman and Corden, 1939; Waksman et al., 1939a; Waksman et al., 1939b). Waksman's work tended to be autoecological with a focus on the role of specific actinomycetes during composting. A significant contribution by Waksman and colleagues was the observation of the successional nature of microbial populations and how the activity of earlier populations promoted or inhibited the development of later populations. This selectivity for subsequent populations is complex and may be based on nutrition, niche filling, the production of antagonistic compounds, or growth-promoting vitamins or cofactors. Compost prepared for growing the mushroom *Agaricus brunnescens* is perhaps the best example of biologically induced selectivity.

A third line of research was synecological, looking at overall microbial activity or activity by broad classes of microbes as affected by various environmental factors. These studies established many of the basic ecological parameters within which practical waste composting could be carried out. Lambert and Davis (1934) observed the effect of temperature and oxygen on composting activity in the field. They concluded that microbial activity was greatest at moderate temperatures in the presence of oxygen and that such conditions were most conducive to subsequent colonization by fungi. Composting refuse, Wiley (1955, 1956, 1957) carried out a thorough set of synecological investigations and produced the first reliable heat and mass balances, including metabolic water production, in the waste composting literature. Schulze (1958) used respiratory quotients to distinguish the role of aerobic vs. anaerobic metabolism under different composting conditions and the effect of moisture content on microbial activity. Jeris and Regan (1973) were able to describe broadly the optimal environmental conditions for microbial activity in composting systems based on empirical observation. Bagstam (1978, 1979) followed the development of populations of nonactinomycete bacteria and actinomycete and fungal populations during composting and the effect of environmental factors such as temperature on population development.

More recent work in waste composting microbiology, which will be discussed in the following sections, has typically been more process oriented and synecological in nature, often based on investigations of field or model systems. This more recent work has led to more logical composting process control and integration of composting process goals with the nature of the microbial populations responsible for composting. Interpretation of microbial investigations can now be made within the context of the overall ecology of composting.

4.2.3 Population Trends During Composting

4.2.3.1 Bacteria

Bacteria are now recognized as the most important group of microorganisms in the first and most active period of composting. Nonactinomycete bacteria recovered from composting materials are listed in Table 1. Strom (1985a,b) identified composting populations in refuse in both small laboratory reactors held at a range of temperatures and in samples derived from commercial composting plants. Bacteria tend to dominate the earliest stages of composting because they can multiply rapidly on simple and readily available substrates, and many can withstand high temperatures and low oxygen tensions. Most

TABLE 1   Nonactinomycete Bacteria Identified in Composting Materials

Bacteria	Ref.
Alcaligenes faecalis	Corominas et al. (1987)
Bacillus brevis	Strom (1985b), Brinton and Droffner (1994)
B. circulans complex	Strom (1985a,b)
B. coagulans type A	Strom (1985a,b)
B. coagulans type B	Rothbaum and Dye (1964), Strom (1985a,b), Fermor et al. (1979)
B. megaterium	Corominas et al. (1987)
B. licheniformis	Gregory et al. (1963), Okafor (1966), Strom (1985a,b)
B. pumilus	Corominas et al. (1987)
B. sphaericus	Rothbaum and Dye (1964), Strom (1985a,b), Brinton and Droffner (1994)
B. stearothermophilus	Niese (1959), Rothbaum and Dye (1964), Hayes (1968), Fermor et al. (1979), Strom (1985a,b)
B. subtilis	Hayes (1968), Niese (1959), Fermor et al. (1979)
Clostridium thermocellum	Henssen (1957)
Clostridium sp.	Waksman et al. (1939a)
Escherichia coli	Brinton and Droffner (1994)
Flavobacterium sp.	Macauley et al. (1993)
Pseudomonas sp.	Hayes (1968), Stanek (1971), Fermor et al. (1979)
Pseudomonas aeruginosa	Brinton and Droffner (1994)
Serratia sp.	Macauley et al. (1993)
Thermus sp.	Fujio and Kume (1991)

solid waste composting systems are managed at temperatures above 55°C to promote pathogen destruction. Systems lacking sufficient heat removal to control temperature tend to achieve temperatures as high as 80°C (Finstein and Morris, 1975; Willson et al., 1980; Finstein et al., 1987c). Such high temperatures are strongly selective for thermophilic bacteria, especially the genus *Bacillus* (Hayes, 1968; Fermor et al., 1979). Strom (1985a,b) found that above 65°C compost populations were often reduced to pure cultures of *B. stearothermophilus*.

Recently Brinton and Droffner (1994) reported that a number of bacterial species that contain pathogenic strains can live or grow in composts at thermophilic temperatures normally considered adequate for pathogen kill (Burge et al., 1978; Finstein et al., 1982). In a large EPA study reported by Goldstein et al. (1988), the most commonly encountered potentially pathogenic bacteria found in compost samples were *Salmonella* sp., *Yersinia enterocolitica,* and toxigenic *Escherichia coli.* Microbial competition is likely to be a major factor in the fate of potential pathogens, but this requires further investigation. In poorly stabilized composts, the remaining available nutrients can support the regrowth of *Salmonella* sp. (Burge et al., 1987). The apparent safety (i.e., lack of reported disease) of compost prepared under current temperature recommendations (Federal Register, 1979) would indicate that while further investigation into pathogen survival is of interest, there should not be undo alarm about the pathogenic risks of properly prepared composts.

As composting systems tend toward greater levels of process control, thermophilic bacteria become the most significant decomposers, because these populations are the most practical ones to manage. Systems designed to optimize the growth of thermophilic bacteria are practical because temperatures high enough to kill pathogens can be reached while still permitting high rates of decomposition. Bacterial stabilization of wastes can proceed rapidly, with most decomposition occurring within the first week (Miller and Finstein, 1985; Harper et al., 1992; Hamelers, 1993). For example, in the high temperature (about 80°C) method, about 10% of volatile solids from sewage sludge could be decomposed in 3 weeks; whereas, under more optimal lower temperatures (50 to 60°C), 40% of sludge volatile solids could be decomposed in about 7 days (Miller and Finstein, 1985).

4.2.3.2 Actinomycetes

Table 2 provides a listing of actinomycetes identified from composting materials. While actinomycetes are themselves bacteria, their ecological niche preferences and role in decomposition make them distinct from other bacteria. Their filamentous morphology, nutritional preferences, and slower growth characteristics make the actinomycetes ecologically intermediate between other bacteria and fungi. There has been a lack of work reported in recent years regarding the role of actinomycetes during composting.

Waksman's great interest in actinomycetes established their diversity in composting materials (Waksman and Hutchings, 1937; Waksman and Cordon, 1939; Waksman et al., 1939a,b). Thermophilic actinomycetes such as *Streptomyces* sp., *Thermomonospora* sp., and *Thermoactinomyces vulgaris*, which can tolerate composting temperatures in the 50°C range, are commonly found, and a smaller number of species can be found in composts with temperatures around 65°C (Lacey, 1973). Actinomycetes tend to prefer moist, highly aerobic conditions and a neutral or slightly alkaline pH (Chen and Griffin, 1966; Lacey, 1973). This can put them at a disadvantage when the density of available substrates is high. Actinomycetes tend to be common in the later stages of composting and can exhibit extensive growth (Bagstam, 1978, 1979; Joshua et al., 1994).

The proportions of actinomycetes to other bacteria in a finished mushroom compost tends to be strongly related to compost quality (Macauley et al., 1993). While this has not been investigated in MSW composts, it could be a useful indicator of maturity. High actinomycete levels would be expected in more nutritionally stable (less putrescible) composts produced under aerobic conditions. It appears that the development of higher actinomycetes populations is an important successional precursor to the later development of fungal populations.

TABLE 2   Actinomycetes Identified in Composting Materials

Bacteria	Ref.
Actinobifida chromogena	Fergus (1964), Lacey (1973)
Microbispora bispora	Henssen (1957)
Micropolyspora faeni	Fergus (1971), Gregory et al. (1963), Lacey (1973)
Nocardia sp.	Lacey (1973), Wood (1984)
Pseudocardia thermophilia	Henssen (1957), Fergus (1971)
Streptomyces rectus	Henssen (1957), Fergus (1971), Hayes (1968), Stanek (1971)
S. thermofuscus	Makawi (1980)
S. thermoviolaceus	Fergus (1971), Stanek (1971)
S. thermovulgaris	Fergus (1971), Hayes (1968), Stanek (1971), Fermor et al. (1979)
S. violaceus-ruber	Fergus (1971)
Streptomyces sp.	Norman (1930), Waksman et al. (1939b), Tendler and Burkholder (1961), Lacey (1973), Wood (1984)
Thermoactinomyces vulgaris	Waksman et al. (1939), Forsyth and Webley (1948), Erikson (1952), Gregory et al. (1963), Fergus (1971), Lacey (1973), Fermor et al. (1979)
T. sacchari	Lacey (1973), Makawi (1980)
Thermomonospora curvata	Fergus (1971), Stanek (1971), Stutzenberger (1971)
T. viridis	Hennsen (1957), Fergus (1971), Lacey (1973), Fermor et al. (1979)
Thermomonospora sp.	Hennsen (1957), Lacey (1971)

4.2.3.3 Fungi

Fungi identified during composting are reported in Table 3. Based on many reports, species of *Penicillium* and *Aspergillus* appear to be the most common fungi in composting materials. Recent work by Straatsma et al. (1994) revealed a vigorous growth of diverse thermophilic fungi at moderately high temperatures during composting for the preparation of mushroom growing compost. Their work demonstrates the successional nature of population development, where the growth of certain organisms such as *Scyatalidium thermophilum* can select for the specific growth of a later population such as *Agaricus brunnescens*. Conceptually, this work is consistent with earlier similar observations made by Waksman and Hutchings (1937).

During the more active stages of waste composting, temperatures over 55°C are usually achieved, discouraging fungal growth (Fermor et al., 1979). Brock (1978) reported that most fungi cannot grow at 50°C and only a few can grow even poorly at 62°C. Kane and Mullins (1973) found that fungi were excluded from the high temperature stages of composting. Fungi are commonly recovered from composting materials later in processing when temperatures are

TABLE 3   Fungi Identified in Composting Materials

Fungus	Ref.
Zygomycetes	
Absidia corymbifera	Chang and Hudson (1967), Straatsma et al. (1994)
A. ramosa	Chang (1967), De Bertoldi et al. (1983)
Absidia sp.	Gregory et al. (1963)
Mortierella turficola	Hayes (1968)
Mucor miehei	Cooney and Emerson (1964), Festenstein et al. (1965)
M. pusillus	Cooney and Emerson (1964), Fergus (1964), Festenstein et al. (1965), Okafor (1966), Chang and Hudson (1967), Hayes (1968), Kane and Mullins (1973), Fermor et al. (1979), De Bertoldi et al. (1983)
Rhizomucor sp.	Okafor (1966)
Ascomycetes	
Allescheria terrestris	De Bertoldi et al. (1983)
Chaetomium thermophilim	Fergus (1964), Chang and Hudson, (1967), Tansey (1981), Kane and Mullins (1973), Fermor et al. (1979), De Bertoldi et al. (1983)
Dactylomyces crustaceious	Kane and Mullins (1973)
Emericella nidulans	Straatsma et al. (1994)
Myriococcum albomyces	Cooney and Emerson (1964)
Talaromyces dupontii	Fergus (1964), Chang and Hudson (1967), DeBertoldi et al. (1983)
T. emersonnii	Tansey (1981), Straatsma et al. (1994)
T. thermophilus	Tansey (1981), De Bertoldi et al. (1983)
Thermoascus aurantiacus	Stutzenberger et al. (1971), De Bertoldi et al. (1983)
T. crustaceus	Straatsma et al. (1994)
Corynascus thermophilus	Fergus and Sinden (1969), Hedger and Hudson (1970), Straatsma et al. (1994)
Thielavia terrestris	Samson et al. (1977)
Basidiomycetes	
Coprinuslagopus	Cooney and Emerson (1964)
Coprinus sp.	Cooney and Emerson (1964), Chang and Hudson (1967)
Coprinus cinereus	Straatsma et al. (1994)
Lenzites sp.	De Bertoldi et al. (1983)

TABLE 3 (continued)   Fungi Identified in Composting Materials

Fungus	Ref.
Deuteromycetes (Hyphomycetes)	
Aspergillus fumigatus	Carlyle and Norman (1941), Fergus (1964), Gregory et al. (1963), Hayes (1968), Henssen (1957), Stutzenberger et al. (1971), Fermov et al. (1979), Tansey (1981), De Bertoldi et al. (1983), Millner et al. (1977)
Humicola grisea	Fergus (1964), Hayes (1968), Tansey (1981), Fermor et al. (1979)
H. insolens	Chang and Hudson (1967), Fergus (1964), Hayes (1968), Fermor et al. (1979), De Bertoldi et al. (1983)
H. lanuginosa	Gregory et al. (1963), Fergus (1964), Festenstein et al. (1965), Okafor (1966), Chang and Hudson (1967), Hayes (1968), Kane and Mullins (1973), Millner et al. (1977), Tansey (1981), De Bertoldi et al. (1983)
Paecilomyces variotii	Straatsma et al. (1994)
Sporotrichum thermophile	Henssen (1957), Chang and Hudson (1967), Hedger and Hudson (1970), Tansey (1981), De Bertoldi et al. (1983)
Malbranchea pulchella	Cooney and Emerson (1964), Chang and Hudson (1967)
Scytalidium thermophilim	De Bertoldi et al. (1983), Straatsma et al. (1994), Hayes (1968), Kane and Mullins (1973), Millner et al. (1977), Ross and Harris (1983a,b)
Mycelia Sterilia	
Papulaspora thermophila	Fergus (1971)
Myriococcum thermophilim	Straatsma et al. (1994)

more moderate and remaining substrates are predominately cellulose and lignin (Chang and Hudson, 1967; Fergus and Sinden, 1969; DeBertoldi et al., 1982).

Fungi have the ability to break down many complex polymers, including lignin and organic chemicals not produced in nature. For example, *Phanerochaete chrysosporium*, which has been of great interest for the decomposition of various complex wastes (Aust and Benson, 1993), grows vigorously under composting conditions at 39°C (Kerem et al., 1992). Hogan et al. (1988) found that the degradation of certain polynuclear aromatics and polychlorinated biphenyls (PCBs) during composting was enhanced at mesophilic temperatures. In composting systems with good environmental control, temperatures favorable to fungal growth are feasible.

4.2.4 Development and Roles of Different Populations

Studies of the development of different populations in composting systems have tended to be limited to observations of changes within the broad classes of nonactinomycete bacteria, actinomycetes, and fungi. Considering the rapid changes that occur during composting, and the large number of potential organisms (Sections 4.2.3.1, 4.2.3.2, and 4.2.3.3), trying to follow population changes of all potential organisms at the genus or species level is not feasible using current techniques.

The dynamics of population change in various types of self-heating eco-systems show some general trends. Mesophilic bacterial populations tend to peak first, followed by thermophilic bacteria, thermophilic actinomycetes, and lastly thermophilic fungi. Mesophilic bacteria populations can peak in a range of 10^9 to 10^{13} per gram wet weight of fresh substrate at temperatures of 45 to 50°C. At temperatures of 55 to 65°C, thermophilic bacterial counts can peak at 10^8 to 10^{12}. Above 65°C, counts of thermophilic bacteria decrease, but enumeration is difficult because most of the thermophiles are spore formers. As temperatures fall below 65°C, actinomycete counts increase to 10^7 to 10^9, about an order of magnitude less than the counts of other bacteria. Fungal populations increase later as overall composting activity declines and temperatures drop below 50°C. Although enumeration of fungi varies between laboratories, the reported range is 10^5 to 10^8 CFU. Fungi normally are absent at compost temperatures above 60°C, and actinomycetes are absent above 70°C.

Temperature is not the only factor affecting the succession of populations in composting ecosystems. Nutrition is a major factor in determining the relative advantage of some populations over others. Bacteria and actinomycetes are favored by C:N ratios on the order of 10:1 to 20:1, while fungi are favored by ratios as high as 150:1 or 200:1 or even higher for some fungi (Griffin, 1985). Such ratios have to be considered within the context of availability, because carbon or nitrogen held in complex polymers can be poorly available. C:N ratios for MSW are optimal in the 35:1 to 50:1 range (Tchobanoglous et al., 1977) . Such ratios are actually favorable to bacteria in that a fair amount of carbon in MSW waste is initially unavailable until further decomposition has occurred. By the end of MSW composting, C:N ratios tend to narrow into the range of 10:1 to 20:1.

Bacteria tend to grow well on proteins and other simple nitrogen containing compounds. Excess nitrogen is primarily released as ammonia at pH 8 and higher or as ammonium ion at pH 7 and lower (Miller et al., 1991). Ammonia release is significant in that it tends to solubilize complex carbon sources, making them available for uptake by microbes (Myers and Thien, 1988). Higher ammonia concentrations (up to about 1000 ppm) are well tolerated by bacteria but inhibitory to many fungi (Miller et al., 1991). Much of the ammonia released by bacteria from proteins can react with complex carbohydrates (Nommik, 1965, 1970) which become preferred nitrogen sources for many fungi (Fermor et al., 1985). Furthermore, nitrogen sequestered in bacterial cells provides a stable but available form of nitrogen for fungi, because many fungi can digest bacterial cells for nitrogen nutrition. Because of the desirability of having an active thermophilic bacteria population early in the composting process, solid wastes lacking available nitrogen can benefit by the addition of high nitrogen-containing wastes such as sewage sludge or manure.

The assimilation of carbon and nitrogen has a role in growth and nutritional uptake. Generally, microbes that tend to grow quickly will assimilate lower proportions of substrate carbon through metabolic inefficiency, while slower growers are more efficient at carbon assimilation. Maximal growth rates of fungi tend to be about an order of magnitude slower than those of bacteria (Griffin, 1985). Nonactinomycete bacteria tend to assimilate 5 to 10% of substrate carbon, while actinomycetes assimilate 15 to 30%, and fungi assimilate 30 to 40% (Alexander, 1977). Actinomycetes and other bacteria have a 5:1 protoplastic C:N ratio, while fungi have a 10:1 ratio. Relating assimilation ratios and protoplasmic C:N requirements, nonactinomycete bacteria need 1 to 2% nitrogen to degrade a unit of carbon, actinomycetes need 3 to 6%, and fungi need 3 to 4% (Alexander, 1977). Nonactinomycete bacterial utilization of nitrogen tends to be poorly conservative (nitrogen is lost as ammonia), while fungi tend to conserve nitrogen, and actinomycetes fall somewhere between other bacteria and fungi.

Much of the carbon in MSW is in the form of lignocellulose. In the later stages of composting, fungi have a nutritional advantage over bacteria in that they generally have a greater ability to degrade such materials (Chang and Hudson, 1967; Bumpus et al., 1985; Fermor, 1993). The succession of non-actinomycete bacteria, actinomycetes, and fungi fits into the pattern of the most readily available substrates being used first, with the most difficult substrates utilized last.

4.3 MICROBIAL ECOLOGY OF COMPOSTING

4.3.1 Composting as a Rapidly Changing System

One of the most unusual attributes of composting ecosystems is the intensity and speed with which changes occur. Managing a composting system is to a large extent managing the consequences of microbial activity so that a favorable composting environment can be maintained. Under conditions that optimize decomposition rates, temperature within a composting mass can increase by over 30°C in less than 24 hours (Harper et al., 1992). Initial moisture contents of 65% can decrease to 20 to 30% in little more than a week (Finstein et al., 1983, 1985, 1986). Free ammonia concentrations can increase to 1500 μg/g within a day and then decrease to less than a tenth of this value in about the same amount of time (Miller et al., 1990). Counts of different species can increase or decrease quite rapidly during the early stages of composting.

Populations of thermophilic bacteria can exhibit log phase growth early in processing, while heat-sensitive populations can be reduced by log values within hours as high temperatures are achieved. The rapid changes that can occur in composting systems are a function of the very high rates of microbial activity possible in such systems. While rapid decomposition is advantageous, the potential for rapid change also makes process management challenging. Most of the advancements in composting in the past 20 years have been related to improvements in process control to keep the rapidly changing composting system operating under favorable conditions.

4.3.2 Composting as a Matric Phase Ecosystem

4.3.2.1 Structure of the Matrix

Composting occurs within a physical matrix that is more ecologically complex than an aqueous ecosystem. The composition and structure of the MSW compost matrix have a profound effect on both biological activity and process control.

A significant trend in recent years is to separate those materials which are not compostable out of the solid waste stream prior to composting. Those fractions that are best separated into the compostable fraction include food waste, paper waste, yard waste, and wood. Many communities now separately collect yard waste, greatly decreasing this fraction in MSW. The fraction of food, yard, and wood waste, with seasonal and regional variation, may account for 20 to 30% by weight of the total waste stream. Another large fraction that can be composted is the total paper stream. The amount of the total paper stream entering into the composting stream will vary regionally based on markets for recovered paper materials and recovery technologies employed. As a result of an increase in source separation, MSW composition is becoming more influenced by community waste management strategies.

The nature of the MSW being composted can be affected by the proportions of the various compostables separated into the composted waste stream. The C:N ratios are fairly narrow for the food and yard waste fractions, while the C:N ratios are quite wide for paper and wood (Tchobanoglous et al., 1977). The removal or retention of paper in the compostable fraction can greatly affect the bulk density and C:N ratio of the material to be composted.

Soil microbial ecology can serve as a basis for understanding composting matrices. Transfers of moisture, gases, and heat have all been well characterized in soil systems, and these models are useful for composting matrices. A big difference between soil and compost matrices is that composting matrices exhibit a higher substrate density per unit volume. In composting systems, the substrate is the matrix. Composting matrices are physically unstable as their very substance tends to be structurally decomposed, becoming softer and more dense.

At a millimeter scale level, the matrix is comprised of particles of materials aggregated to various degrees, and these aggregates are spatially associated with other aggregates. Particles or aggregates of particles offer a wide range of ecological niches at the microbial level. Within an aggregate, conditions such as temperature and available water will be fairly consistent. Within particles, there are gradients of oxygen and carbon dioxide which can only diffuse slowly when water is present (Hamelers, 1993). These gradients can result in anaerobic conditions at the center of the particle, while conditions can be fully aerobic toward the outer edges. Substrate availability can also vary within a particle. Such gradients provide a large variety of nonequivalent microsites which foster greater species diversity (Wimpenny et al., 1984).

At the decimeter scale level, there can be wide gradients of temperatures, interstitial gas concentrations, moisture contents, pH, and other factors. These physical and chemical gradients result in microhabitats which influence microbial activity (Miller, 1984, 1989). Transfer of gases, heat, water, nutrients, and microbial populations lead to the development of distinct gradients. A great deal of process control in waste composting involves overcoming the potential for large-scale gradient formation and trying to maintain favorable environmental conditions within much of the compost.

4.3.2.2 Effect of the Matrix on Phenotypic Expression

Association with a matrix tends to have a range of effects on microorganisms including the stabilization of membranes and the induction of different types of phenotypic expression. For example, Filip (1978) found that the association of microbes with particles in composting refuse allowed them to tolerate greater heat stress. Referring to phenotypic expression, an obvious example is the dimorphic fungi which grow as yeast in liquid media but as mycelia on solid substrates (Alexopoulous and Mims, 1979). Another example is the observation by Ohno et al. (1992) that the production of the antifungal antibiotic iturin by *Bacillus subtilis* is 5 to 6 times higher per unit weight of wet substrate on a solid matrix than in submerged culture. Furthermore, the physical structure of the matrix affects colonization of the substrate. Especially for bacteria, cells need to be transported or transferred within the matrix to the available substrate. Using an electron microscope, Atkey and Wood (1983) observed heterogeneity in the colonization of straw particles, with much of the still remaining straw uncolonized at the end of a composting process.

Phenotypic expression of enzymes involved in decomposition is greatly stimulated in many organisms by association with a physical matrix. Ecologically, there is an advantage to expressing certain degradative enzymes in a matrix rather than in an aqueous system where enzymes would diffuse away before benefiting the organism. Most research on the phenotypic expression of enzymes has been carried out with fungi, but this should not be interpreted as evidence that matric associations do not affect phenotypic expression of enzyme production in bacteria and actinomycetes.

Greater expression of enzyme production in matric phase systems include xylanases by *Scytalidium* (Da Silva et al., 1994), *Chaetomium* (Wiacek-Zychlinska et al., 1994), *Thermomyces* and *Thermoascus* (Alam et al., 1994), and

Trichoderma (Haapala et al., 1994). Matric culture also significantly increases the production of amylase by *Aspergillus* (Sudo et al., 1994); amylase and proteinase production by mixed cultures of *Saccharomyces, Schwanniomyces, Yarrowia, Aspergillus* and *Rhizopus* (Yang et al., 1993); and pectinase production by *Aspergillus* (Sudo et al., 1994). Protease production by *Rhizopus* has been demonstrated to be optimal in matric phase cultures with initial moisture contents of 47% (Ikarsari and Mitchell, 1994). For the entire class of brown rot fungi that degrades cellulose, the endoglucanase enzyme system is expressed in matric culture but cannot be demonstrated to function in liquid culture (Kleman-Leyer et al., 1994). Matric effects on the phenotypic expression of enzymes by bacteria and actinomycetes would further our understanding of composting microbiology, but research in this area is currently lacking.

4.3.3 Interactions Between Microbial Activity and the Physical and Chemical Environment

Substrate availability is affected by factors both intrinsic and extrinsic to the material being composted. Intrinsic factors are chemical and structural characteristics of the waste that make it biologically accessible. Lignin intrinsically limits substrate availability in that it can physically block access to other more degradable polysaccharides. Extrinsic factors are those ecological factors that affect the populations that utilize a specific substrate. High temperature is an extrinsic factor which could preclude the growth of fungi that might otherwise attack lignin. Other extrinsic factors related to availability include particle size. Substrate availability can be increased by size reduction through grinding or shredding (Gray and Sherman, 1969). The most reliable measure of substrate availability is the actual measure of substrate utilization during composting (Iiyama et al., 1994).

In lignocellulosic wastes, microbial activity during the most active stage of composting is based on the metabolic utilization of cellulose and noncellulosic polysaccharides (Macauley et al., 1993; Nakasaki et al., 1994). Lignin appears to be structurally altered by significant oxidation during the thermophilic stages of mushroom composting, but it is not decomposed (Iiyama et al., 1994). Rates of protein decomposition are high early in composting, based on the release of large amounts of ammonia (Miller et al., 1990, 1991). Microbial production of new protein is also significant, so that the overall percentage of protein in the compost tends to stay constant (Macauley et al., 1993). Based on evidence from the composting of grease trap wastes, lipids can be utilized rapidly by bacteria, including actinomycetes, under thermophilic conditions (Joshua et al., 1994). The decomposition of sugars, polysaccharides, proteins, and lipids are within the general metabolic capability of species of *Bacillus* and *Clostridium* (Brock and Madigan, 1988).

As heat and gas transfer rates per unit volume of compost are consistent and low, the major changeable variable determining composting temperature and gas concentrations is the available "energy density". Energy density refers to the metabolically available food energy per unit volume of substrate. Examples of a low-energy density substrate would be straw, while grease trap waste is a high-energy density substrate. Solid waste general has a moderate energy density. Temperature and gas concentrations within a composting mass are a function of (1) the rate of metabolic heat liberation, (2) the uptake of oxygen and release of other gases, and (3) the rate of exchange of heat and gases with the ambient atmosphere. Since the rates of both heat and gas transfer to the ambient atmosphere are a function of distance, the physical volume of a composting mass provides a fixed limitation on the potential passive rate of heat and gas transfer. Thermal conductivities and gas transfer rates are uniformly low across a wide range of composting materials (Finstein et al., 1986; Miller et al., 1989).

Operationally, the concept of energy density provides a framework for the consideration of the most appropriate means of controlling a composting process. While a turned windrow method might work well for a low energy density substrate, forced ventilative control of heat and gas transfer might be required to control a high-energy density substrate. Energy density can be measured directly via heat modeling of substrates as they are composting (Harper et al., 1992; Miller, 1984; Hogan et al., 1989). Energy density can be modified by the mixing of low-energy density substrates such as paper into high-energy density substrates such as food waste. Physical compaction also can affect energy density by increasing the substrate bulk density by decreasing void space.

4.3.3.1 Heat Evolution

Heat evolution during composting is caused by microbial activity; only under rare conditions would chemical reactions make any contribution to heat production (Nell and Wiechers, 1978). In large composting masses, temperatures do not exceed about 80°C, which is also the temperature at which biological activity effectively stops (Miller et al., 1989). Heat evolution is of practical importance for temperature and moisture management. Heat evolution can be calculated based on heats of combustion (Haug, 1979) or measured directly. A difficulty with calculated values is that they do not account for the selective nature of metabolism (that is, some substrates are preferred over others) or the production of new metabolic products with different levels of chemical energy.

Composting refuse with a significant grease fraction (14.7%), (Wiley, 1955), measured heat outputs of 22.1 to 28.5 kJ/g decomposed. Miller (1984) reported two data sets for composting primary sewage sludge which averaged 21.8 and 15.2 kJ per gram decomposed. Composting rice hulls and rice flour, Hogan et al. (1989) measured heats per gram decomposed of 14.2 and 16.7 kJ in various trials. In 10 trials with a mushroom composting substrate (mainly straw and chicken litter) Harper et al. (1992) measured heats per gram decomposed ranging from 15.4 to 22.0 kJ, with an average value of 18.3 kJ. Studies measuring heat evolution demonstrate that measured values are lower than calculated values.

Oxygen uptake is related to heat evolution. Calculation based on combustion gives a value of 14 MJ released per kg oxygen consumed (Finstein et al., 1986). Harper et al. (1992), reporting on hundreds of observations with a mainly straw and poultry mixture, found 9.8 MJ were released per kg oxygen consumed. This difference of 4.2 MJ can be accounted for partly by the greater level of oxidation of the final compost residue compared to the initial ingredients.

Practical temperature control can only be achieved by heat removal based on evaporative cooling and sensible heating of dry air (Kuter et al., 1985; Finstein et al., 1986). Turning compost removes large amounts of heat as the hot compost is moved through the air (Miller et al., 1989). For materials not turned, forced pressure (or the less efficient vacuum pressure) ventilation can remove great amounts of heat (Miller et al., 1982). Macgregor et al. (1981) calculated that for large composting masses approximately 90% of the heat removed through ventilation was through evaporative cooling, and only 10% was due to sensible heating of the dry air. This calculation was based on air inlet conditions of 60% relative humidity (RH) at 25°C and exit conditions of 100% RH at 50 to 70°C. Evaporation is the most important mechanism for heat removal because of the high heat of vaporization of water. Ventilative heat removal to control temperature is best implemented through systems that remove heat through temperature feedback controlled ventilation (De Bertoldi et al., 1982; Kuter et al., 1985; Finstein et al., 1986). Temperature feedback control is now widely accepted for ventilated systems.

4.3.3.2 Temperature

Temperature is the most critical factor to control during composting because of the effect of temperature on microbial metabolic rates and population structure. For any individual microbial population, growth rates increase rapidly with temperature, but decline rapidly after the temperature optimum is reached (Zwietering et al., 1991). If the temperature of a composting mass is much below 20°C, microbial activity is low, and an appreciable lag period may occur. As temperature exceeds 20°C, microbial activity increases, since enzyme activity rates generally double with each 10°C rise in temperature (Atlas and Bartha, 1981). As composting proceeds, mesophilic populations are inhibited by temperatures exceeding the lower 40s, while in the mid-40s conditions become favorable for the growth of thermophiles. Above 60°C, temperature starts to inhibit microbial activity as the temperature optimum for various thermophiles is exceeded. Optimal decomposition rates typically are in the range of 55 to 59°C. Optimal and suboptimal temperature ranges have been identified by various methods including measurement of microbial heat production and mass balance measures of decomposition.

Achievement of high temperatures is desired not only to speed decomposition rates but also to kill pathogens. Some flexibility in controlling composting temperatures is possible in that, even if the optimal temperature range of some organisms is exceeded or not reached, there is a diversity of organisms that can still permit the overall process to succeed. The maximum temperature achievable through composting is approximately 82°C, at which point biological activity and metabolic heat evolution cease (Finstein et al., 1986; Nell and Wiechers, 1978). This ceiling value of 82°C also has been observed in large-scale field operations (Finstein et al., 1983) and is commonly observed in commercial practice. While extreme thermophiles exist, in deep sea vents for instance, they can only live in low densities because of oxygen limitations related to temperature–solubility functions (Sundaram, 1986). McKinley and Vestal (1984) found no evidence that extreme thermophiles contributed to decomposition during composting. Strom (1985b) identified *Bacillus stearo-thermophilus* as the most extreme thermophile isolated from a composting ecosystem.

4.3.3.3 Oxygen and Gas Exchange

Gas exchange is crucial during composting because oxygen-based microbial metabolism is more efficient than fermentation. A lack of oxygen is a common reason for composting failures (Gray et al., 1971; Poincelot, 1975). Oxygen concentrations are determined by rates of diffusion and microbial uptake. Total diffusion rates are determined by the gas concentration gradient and the resistance to flow. In turn, flow resistance is a function of pore size, pore continuity, and water film thickness. Within the gas-filled interstitial atmosphere, diffusion rates have been measured for refuse (McCauley and Shell, 1956) and for sewage sludge cakes (Nakasaki et al., 1987). Passive diffusion within the interstitial atmosphere is much too slow to supply a large composting mass with sufficient oxygen; therefore, active ventilation is required.

As water content increases, gas diffusion becomes restricted until metabolic demands exceed supply. The coefficient of oxygen diffusion through gas-filled space is 0.189 cm²/sec (Letey et al., 1967), but only 2.56×10^{-5} cm²/sec through water (Baver et al., 1972), or almost 10,000 times slower. Gaseous flow potential is best determined by the cross-sectional void space (Papendick and Campbell, 1981). For solid waste composting systems, requirements for free air space of 30 to 35% of the total volume have been proposed for the maintenance of sufficient gas diffusion (Schulze, 1961; Haug, 1978). Various investigators have suggested that aerobic activity can be maintained with interstitial oxygen concentrations of 5% (Schulze, 1962; Parr et al., 1978; Willson et al., 1980), 10% (Suler and Finstein, 1977), or 12 to 14% (Harper et al., 1992; Miller et al., 1990). Miller (1984) found that the rate of compost decomposition was increased by 25% in constant temperature trials where interstitial oxygen concentrations were maintained above 18%, but ventilation was doubled to improve diffusion. De Bertoldi et al. (1988) reported composting activity rates were highest when interstitial oxygen concentrations were maintained between 15 and 20%.

As for soil particles (Greenwood, 1961; Foster, 1988), composting matrices can contain aerobic and anaerobic microenvironments coexisting within close proximity, or larger zones can be aerobic or anaerobic (Miller et al., 1989). Microbial activity can be retarded by oxygen limitations even with seemingly high interstitial oxygen concentrations. When interstitial oxygen is present, products of anaerobic metabolism (such as H_2S or CH_4) can still be recovered from composting materials (Op den Camp, 1989). The rapid consumption of oxygen along with the matric nature of the solid waste substrate means that even in an "aerobic" compost there can be significant anaerobic metabolism occurring within particles.

If the composting matrix is allowed to become predominately anaerobic, volatile organic acids (VFAs), along with other reduced sulfur and nitrogen compounds can accumulate (Miller and Macauley, 1989). Composting under anaerobic conditions can result in the conversion of up to 15% of the total organic carbon content into VFAs (Chanyasak et al., 1980). The rate of formation of VFAs under optimal aerobic conditions can be even greater than the formation under anaerobic conditions, but under aerobic conditions the pH remains high enough that the VFAs are present in ionic form (Cerny, 1992). The ionic forms of the VFAs are much less toxic than the free acid forms. While the free VFAs are quite volatile and odorous, the ionic forms are highly water soluble and not volatile. Ionic forms of VFAs are good substrates for microbial metabolism under aerobic conditions. VFAs produced under anaerobic conditions might be broken down under subsequent aerobic conditions, but they sometimes appear to persist at inhibitory concentrations. Organic acids are toxic to higher plants and overly anaerobic composts can remain phytotoxic for years (Hoitink, 1980).

4.3.3.4 Moisture

Microorganisms have physiological requirements for water; water is also crucial for the transport of many microbes (Griffin, 1981a,b; Harris, 1981; Miller, 1989). As a general rule, optimal microbial activity is achieved by the maximum water content that does not restrict oxygen utilization. While the use of water activity for composting investigations has been proposed (Finstein and Morris, 1975), water activity is much too insensitive to be of use. An alternative is the measure of water potential which separates associations lowering the free energy of water including the osmotic and matric potentials. Osmotic potential describes the lowering of water free energy through association with dissolved compounds, while matric potential describes the lowering of water free energy through the interaction of water with capillaries and surfaces. Water potentials are energy relationships in reference to free water and are expressed in Pascals (Pa).

Matric potential is of ecological significance as a basis for predicting microbial colonization and overall activity. Matric potential can explain activity limitation caused by dryness. As matric potentials decrease below −70 kPa, colonization by bacteria becomes strongly inhibited because of a lack of adequate water for cell transport (Miller, 1989). Such limitations have been demonstrated for soil matrices (Griffin and Quail, 1968). Bacterial colonization is most favored at matric potentials above approximately −20 kPa (Miller, 1989). Fungi and actinomycetes are less affected by matric potential because they can colonize via hyphae across air gaps. Figure 2 provides a visual illustration of the effect of matric water potential on microbial development.

Colonization limitations due to dryness can be somewhat alleviated by mixing to aid in microbial dispersal. Continuously mixed systems can exhibit high levels of activity even though the composting material is fairly dry (Shell and Boyd, 1969). A well-colonized compost kept under otherwise favorable conditions still can be active at moisture contents as low as 22% (Finstein et al., 1983). Wastes with a high proportion of readily available substrate will release sufficient heat to drive considerable water removal, and final moisture contents as low as 20 to 30% can be achieved (Miller et al., 1980; DeBertoldi et al., 1982; Miller and Finstein, 1985). With some wastes, so much drying can occur that water can become rate limiting, and addition of water will renew activity and lead to a more mature compost (Macgregor et al., 1981; Robinson and Stentiford, 1993). Another consequence of microbial activity is the production of metabolic water. Haug (1979) used combustion model values to calculated that 0.72 g of water should be produced for every gram of sewage sludge decomposed. Actual measured values for metabolic water production are consistently lower, in the range of 0.50 to 0.55 g of water produced (Hogan et al., 1989).

4.3.3.5 pH and Ammonia

Values of pH below 6 can retard initial composting activity, although this normally will not be inhibitory once composting is well under way (Schulze, 1962; Jeris and Regan, 1973b). Preventing pH from dropping below 7 by lime addition has been shown to permit faster degradation of refuse, and the degradation of proteins by microorganisms is fastest in the range of pH 7 to 8 (Nakasaki et al., 1993). Problems related to high pH in MSW composting have not been reported.

Prediction of the extent of ammonia release can be made based on the C:N ratio (Witter and Lopez-Real, 1987). Early in the composting process, microbial ammonification will cause pH to rise into the 8.0 to 8.5 range. As readily available nitrogen-containing substrates are used up, ammonification will decline and pH will drop back into the 7.5 to 8.0 range, usually within a week. Even in a mature compost, the pH normally will be somewhat above 7. Ammonification occurs most rapidly between 40 and 50°C (Ross and Harris, 1982), but the greatest conservation of ammonia nitrogen is between 55 and

FIGURE 2   A laboratory reactor for the study of composting of sewage sludge and wood chips. At the bottom of the reactor, a dense growth of fungal mycelium is visible. In the middle of the reactor, chalky white actinomycete colonies may be seen. (Photograph copyright Frederick C. Miller.)

50°C (Burrows, 1951). Because pH is linked to the decomposition of readily available nitrogen compounds, the course of pH change can be used as an indicator of microbial activity.

Since the conservation of nitrogen in the compost by conversion into less available forms is the desired outcome in MSW composting, attention to the role of pH in ammonia dissociation is warranted. While pH is affected by ammonification, pH has a great effect on the ammonia (NH_3)–ammonium ion (NH_4^+) equilibrium. Significantly, the chemical and biological activity of ammonia and ammonium ion are very different. The relationship between pH and ionization of ammonia can be described by the following equation (Koster, 1986):

$$\text{Fraction free } NH_3 = [1 + \log(pK_w - pK_b - pH)]^{-1}$$

Values for the dissociation constant of water (K_w) and the ionization constant for ammonia (K_b) at 30°C would be $pK_w = 13.833$ and $pK_b = 4.740$. This relationship is such that at pH 7.0 and below, the ammonium ion form is dominant, while above pH 10 free ammonia dominates.

Free ammonia is highly reactive and will chemically combine with organic matter, including carboxyls, carbonyls, enolic, phenolic and quinone hydroxyls, and unsaturated carbons near functional groups (Nommick, 1965; Sharon, 1965; Mortland and Walcott, 1965). Reaction rates increase with temperature (Nommick, 1970) by the presence of oxygen (Mortland and Walcott, 1965). Free ammonia has been observed to dissolve organic matter and make it susceptible to further decomposition (Myers and Thien, 1988).

4.3.3.6 Substrate Changes

During composting the most readily available substrates are degraded, and a residue of more resistant materials remains. By analogy, changes that occur during composting are not dissimilar to the substrate changes that occur to organic materials incorporated into soil, with the exception that composting occurs on a much shorter time scale. Decomposition in soils progresses through a pathway described as humification (Stevenson, 1982; Tate, 1987). Generally, more bioavailable organic matter is slowly mineralized, while the remaining organic matter tends to become more highly cross linked and oxidized. Composted solid waste not only becomes more humified over time, but the degree of humification can be used as an indicator of compost maturity (Ciavatta et al., 1993).

Macauley et al. (1993) reported the results of a biochemical analysis of the changes occurring with time in a mushroom compost composed of mainly straw and poultry manure. The nature of changes observed is relevant to MSW composting as the major substrate is lignocellulose in both cases. Comparing the starting mixture to the compost at the end of phase I (equivalent to the beginning of curing in waste compost), the greatest relative loss occurred within the noncellulosic polysaccharides fraction, followed by the cellulosic fraction, the water soluble fraction, the 80% ethanol-soluble fraction, and the phenolic acid fraction. There was little or no change in the protein fraction, reflecting that while much protein was being broken down, significant production of new protein occurred. There was no change in the lignin fraction, which appears to be recalcitrant under composting conditions.

4.4 FORMATION OF ODOROUS COMPOUNDS

The nature and formation of composting odors have recently been reviewed by Miller (1992) and Kissel et al. (1992). Table 4 lists some compounds frequently implicated in odor problems. The microbiology of composting odors has been considered synecologically, but specific autoecological investigations are lacking. High temperatures during the most active early stage of composting restricts microbial populations almost exclusively to the endospore producing bacteria (see Section 4.2.3) in the genera *Bacillus* and *Clostridium* (Strom, 1985a,b). *Bacillus* are aerobes or facultative aerobes, while *Clostridium* are anaerobes. Investigation of anaerobic bacteria is often avoided because of technical difficulties in culturing such organisms (Brock and Madigan, 1988); culture work with thermophilic bacteria is also technically difficult and requires special methods (Edwards, 1990). The difficulties of working with thermophilic anaerobes have prevented direct investigation of microbial formation of odorous compounds under composting conditions.

Bacillus and *Clostridium* have the fermentative ability to produce a broad range of odorous compounds (Brock and Madigan, 1988). Anaerobic and thermophilic (up to 65°C) sulfate reduction to hydrogen sulfide in organic rich substrates can be carried out by *Desulfotomaculum thermobensoicum* and *Desulfotomaculum orientis* under laboratory conditions (Tanimoto and Bak, 1994). While there are no reports of recovery of *Desulfotomaculum* from composting materials, the ecological requirements of these organisms would indicate their presence in composting materials is likely. Sulfate reduction to hydrogen sulfide occurs commonly in high temperature zones of composting materials when sulfate is present (Miller et al., 1991).

The majority of compounds causing composting odors are formed directly via microbial activity. Notable exceptions are carbon disulfide and dimethyl disulfide, which can be produced nonbiologically in increasingly large amounts as compost temperatures exceed the 60 to 70°C range (Miller and Macauley, 1988; Derikx et al., 1990). Reduced sulfur and nitrogen compounds and organic acids are principal causes of odor problems and are common products of anaerobic activity. Redox potential will determine the types of odorous compounds formed, such as dimethyl disulfide production peaking at 0 mV

TABLE 4   Compounds Contributing to Composting Nuisance Odors

Compound	Chemical Formula	Molecular Weight	Boiling Point	Odor
Sulfur-Containing Compounds				
Hydrogen sulfide	H_2S	34.1	–60.7	Rotten egg
Carbon oxysulfide	COS	60.1	–50.2	Pungent
Carbon disulfide	CS_2	76.1	46.3	Sweet
Dimethyl sulfide	$(CH_3)_2S$	62.1	37.3	Rotten cabbage
Dimethyl disulfide	$(CH_3)_2S_2$	94.2	109.7	Sulfide
Dimethyl trisulfide	$(CH_3)S_3$	126.2	165	Sulfide
Methanethiol	CH_3SH	48.1	6.2	Sulfide, pungent
Ethanethiol	CH_3CH_2SH	62.1	35	Sulfide, earthy
Ammonia- and Nitrogen-Containing Compounds				
Ammonia	NH_3	17.0	–33.4	Pungent, sharp
Aminomethane	$(CH_3)NH_2$	31.6	–6.3	Fishy, pungent
Dimethylamine	$(CH_3)_2NH$	45.1	7.4	Fishy, amine
Trimethylamine	$(CH_3)_3N$	59.1	2.9	Fishy, pungent
3-Methylindole (skatole)	$C_6H_5C(CH_3)CHNH$	131.2	265	Feces, chocolate
Volatile Fatty Acids				
Methanoic (formic)	HCOOH	46.0	100.5	Biting
Ethanoic (acetic)	CH_3COOH	60.1	118	Vinegar
Propanoic (propionic)	CH_3CH_2COOH	74.1	141	Rancid, pungent
Butanoic (butyric)	$CH_3(CH_2)_2COOH$	88.1	164	Rancid
Pentanoic (valeric)	$CH_3(CH_2)_3COOH$	102.1	187	Unpleasant
3-Methylbutanoic (isovaleric)	$CH_3CH_2CH(CH_3)COOH$	102.1	176	Rancid cheese
Ketones				
Propane (acetone)	CH_3COCH_3	58.1	56.2	Sweet, minty
Butanone	$CH_3COCH_2CH_3$	72.1	79.6	Sweet, acetone
2-Pentanone	$CH_2COCH_2CH_2CH_3$	86.1	102	Sweet
Other Compounds				
Benzothiozole	C_6H_4SCHN	135.2	231	Penetrating
Ethanol	CH_3OH	44.1	20.8	Sweet
Phenol	C_6H_5OH	94.1	181.8	Medicinal

Source: Modified from Miller, F.C. 1992. Minimizing odor generation, in H.A.J. Hoitink and H.M. Keener (Eds.), *Science and Engineering of Composting: Design, Environmental, Microbiological and Utilization Aspects,* Renaissance Publications, Worthington, OH, pp. 219–241.

or H$_2$S formation peaking at -100 mV (Beard and Guenzi, 1983). The most anaerobic conditions might not be the worst for odor production, however, because H$_2$S production can decrease with a decrease in solution SO$_4^{2-}$ below -200 mV. Some odorous reduced compounds can be produced under aerobic conditions, such as methanethiol from the decomposition of methionine (Banwart and Bremner, 1975).

Higher temperatures have been observed to lead to greater odor release (Macgregor et al., 1981). One explanation for this is that many of the reduced odorous compounds can be oxidized biologically under aerobic conditions, but the reduction of microbial activity at high temperatures reduces the potential for metabolism of these compounds before they escape. Higher temperatures also can increase the vapor pressure of many odorous compounds leading to higher release rates.

4.5 PROCESS MANAGEMENT

4.5.1 Process Strategy

Process strategy deals with how the composting ecosystem is managed to achieve a selected outcome. Engineering practice tends to define processes by configuration (Haug, 1993), but from an ecological viewpoint microorganisms respond to selective environmental conditions. A common reason for failures in MSW composting has been a lack of a microbially based processing strategy and a configuration to support that strategy. Processing strategy must consider temperature management. Minimal temperatures over a defined time period are important for many compost uses to ensure the killing of pathogens (Burge et al., 1978). Moreover, final moisture content is a function of process strategy. Processing strategy is also needed to achieve the level of compost maturity required for different uses.

4.5.2 Process Configuration

Process configuration is the physical means of handling the composting material. Some configurations permit little or no control over the composting ecosystem, making it impossible to implement a control strategy. Most composting configurations are designed for batch processing, but there are continuous feed systems. Both have advantages and disadvantages, but batch systems are more amenable to precise process control.

4.5.2.1 Windrows

Windrows are long rows of material with a somewhat triangular cross section. Turning of windrows is usually needed for MSW composting. Windrow systems are relatively inexpensive to set up and are suitable for long-term or temporary applications; they can be fully exposed outdoors, covered with a roof, or fully enclosed. The advantage of turned windrows is that the turning operation physically breaks up and mixes the MSW without the need for additional physical processing. This mechanical action is worthwhile, as it reduces the size of larger particles during processing, allowing density to be increased as activity decreases. Turned windrow systems can flexibly handle the mixing of MSW with additional waste streams and facilitate water addition for moisture control. Turned windrows require space between windrows to accommodate turning equipment.

Process control in windrows is limited, and processing conditions can be highly variable spatially. As significant portions of the windrow will be under suboptimal conditions for temperature and oxygen at any one time, processing times are longer than that of more controlled methods.

4.5.2.2 Aerated Windrows

Aerated windrows are built over a ventilation system that greatly aids in the control of windrow temperature and oxygen concentration. Ventilation can be in vacuum or pressure mode. Generally, forced pressure is more common (Goldstein and Steuteville, 1993) as it is more mechanically efficient and provides better process control (Miller et al., 1982). Some composting facilities operate in a vacuum mode so that exhaust gases can be directed through an odor scrubbing system (Forbes et al., 1994).

4.5.2.3 Aerated Static Piles

The aerated static pile is the most popular method for the composting of wastes, other than yard waste, in the U.S., accounting for half of the total number of waste composting facilities (Goldstein and Steuteville, 1993). Most of these facilities, however, are being used for the composting of biosolids. Aerated static pile composting by itself is not well suited to the composting of MSW because of the lack of mixing. This problem can be overcome, however, by the use of a mixed composting process stage prior to aerated static pile (Steuteville, R., personal communication). Space requirements for aerated static pile systems are less than for windrows, because static piles can be "piggy backed" or placed directly against one another. Each row normally has an independent ventilation control system.

4.5.2.4 Bins, Troughs, Enclosed Reactors, and Silos

Bins are a simple form of containment in which something such as the aerated static pile method is carried out but with the compost contained by walls. Advantages of bins are that materials handling is improved, and ventilation management is made easier because the spatial relationship of the composting mass over the ventilation system is consistent. Trough or trench systems are similar to bin systems, except that materials handling is carried out by customized equipment that can run on rails the length of the trough. Trough systems are well suited to MSW since this configuration promotes good mixing and facilitates water addition. Temperature feedback by controlled ventilation is feasible in trough systems, one section at a time.

Enclosed reactors are produced by various companies, while others are custom designed and constructed. Most such systems are designed to automate materials handling and give consistent process control. Many enclosed systems are designed to increase mixing and materials preparation during the earliest phase of composting. Other systems are designed to carry out composting to a fairly high degree of maturity. Enclosed systems can be quite expensive; if improperly designed in reference to process management, they can be impossible to correct.

Silos are an example of a configuration that makes the implementation of temperature control very difficult or impossible. Heat can be transferred from the compost mass to the ventilation air only if the compost is hotter than the air, leading to ever higher composting temperatures as height increases above the ventilation inlet. Depending on the available energy density for a particular solid waste, gradients can increase from 5 to 15°C/m along the ventilation path (Finstein et al., 1986), making high silos uncontrollable. Although interstitial oxygen can be provided all the way to the top of a silo, at temperatures above 80°C there is no longer a microbial metabolic demand for oxygen.

4.5.2.5 Tunnels

Tunnels are an enclosed configuration which permit a very high level of process control not possible in other configurations. In the tunnel configuration, air is used to provide oxygen and remove heat and is used additionally as a heat exchange fluid (Miller, 1990). A great proportion of the air moved in a tunnel is recirculated to distribute heat evenly throughout the composting mass. A smaller fraction of fresh air is introduced into the ventilation stream to provide evaporative cooling and to resupply oxygen. The use of air as a heat exchange fluid overcomes the intrinsic temperature gradient that occurs with single pass-through ventilation. The ability to distribute heat evenly throughout the composting mass permits tight control of temperature throughout the composting mass regardless of the available energy density of the composting substrate. Using fresh air for cooling, interstitial oxygen concentrations of about 18% are continuously maintained, and, coupled with velocity-induced air turbulence, these systems tend to be highly aerobic (Miller et al., 1990). In application, materials with a very high available energy density can be controlled to within a few degrees anywhere within the composting mass. Maintenance of highly aerobic conditions in tunnel composting systems appears to reduce odor production during the preparation of mushroom composts (Overstijns, 1994) and in composting of grease trap waste (Macauley, B.J., personal communication).

4.6 CONCLUSIONS

Composting is a powerful process for the degradation of waste materials. Composting physical ecology has now been fairly well described, and the control of moisture, temperature, gas exchange, and other physical factors can now be practically carried out in full-scale facilities.

While there is a fair idea of which microorganisms are active in composting materials at the different stages of composting, much less is understood about the function and role of individual species. Compositional changes occurring during composting provide insight into biochemical reactions mediated by specific populations and a possible key to understanding population development (Iiyama et al., 1994). With a more complete knowledge of the manner in which physical and nutritional factors determine population development, better control over the decomposition of specific wastes and the formation of useful final compost products will be possible. It is hoped that more modern and rapid means of microbial identification, coupled with a better appreciation of the ecologically selective nature of better controlled composting systems, will permit more definitive investigations of composting population dynamics in the future.

REFERENCES

Alam, M., I. Gomes, G. Mohiuddin, and M.M. Hoq. 1994. Production and characterization of thermostable xylanases by *Thermomyces lanuginosus* and *Thermoascus aurantiacus*. *Enzyme Microb. Tech.* 16:298–306.

Alexander, M. 1977. *Introduction to Soil Microbiology*, John Wiley & Sons, New York.

Alexopoulos, C. J., and C.W. Mims. 1979. *Introductory Mycology*, 3rd ed., John Wiley & Sons, New York.

Atkey, P.T., and D.A. Wood. 1983. An electron microscope study of wheat straw composted as a substrate for the cultivation of the edible mushroom (*Agaricus bisporus*). *J. Appl. Bacteriol.* 55:293–304.

Atlas, R.M., and R. Bartha. 1981. *Microbial Ecology: Fundamentals and Applications*. Addison-Wesley, Reading, MA.

Aust, S.D., and J.T. Benson. 1993. The fungus among us: use of white rot fungi to biodegrade environmental pollutant. *Environ. Health Perspect.* 10:232–233.

Bagstam, G. 1978. Population changes in microorganisms during composting of spruce-bark. I. Influence of temperature control. *Eur. J. Appl. Microbiol. Biotech.* 5:315–330.

Bagstam, G. 1979. Population changes in microorganisms during composting of spruce-bark. II. Mesophilic and thermophilic microorganisms during controlled composting. *Eur. J. Appl. Microbiol. Biotech.* 6:279–288.

Banwart, W.L, and J.M. Bremner. 1975. Identification of sulfur gases evolved from animal manures. *J. Environ. Qual.* 4:363–366.

Bardos, R.P., and J.M. Lopez-Real. 1988. The composting process: susceptible feedstocks, temperature, microbiology, sanitation and decomposition. European Community Workshop on Compost Process in Waste Management, Sept. 1988, Neresheim, Germany.

Baver, C.A., W.H. Gardner, and W.R. Gardner. 1972. *Soil Physics*, 4th ed., John Wiley & Sons, New York.

Beard, W.E., and W.D. Guenzi. 1983. Volatile sulfur compounds from a redox-controlled cattle manure slurry. *J. Environ. Qual.* 12:113–116.

Brinton, W.F., and M.W. Droffner. 1994. Microbial approaches to characterization of composting processes. *Compost Sci. Utiliz.* 2:12–17.

Brock, T.D., and M.T. Madigan. 1988. *Biology of Microorganisms*, 5th ed., Prentice Hall, Englewood Cliffs, N.J.

Brock, T.D. 1978. *Thermophilic Microorganisms and Life at High Temperatures*, Springer-Verlag, New York.

Browne, C.A. 1933. The spontaneous heating and ignition of hay and other agricultural products. *Science* 77:223–229.

Bumpus, J.A., M. Tein, D. Wright, and S.D. Aust. 1985. Oxidation of persistent environmental pollutants by a white rot fungus. *Science* 228:1434–1436.

Burge, W.D., W.N. Cramer, and E. Epstein. 1978. Destruction of pathogens in sewage sludge by composting. *Trans. ASAE* 510–514.

Burge, W.D., N.K. Enkiri, and D. Hussong. 1987. *Salmonella* regrowth in compost as influenced by substrate. *Microb. Ecol.* 14:243–253.

Burrows, S. 1951. The chemistry of mushroom composts. II. Nitrogen changes during the composting and cropping processes. *J. Sci. Food Agric.* 2:403–410.

Carlyle, R.E., and A.G. Norman. 1941. Microbial thermogenesis in the decomposition of plant materials. Part II. Factors involved. *J. Bacteriol.* 4:699–724.

Cerny, R. 1992. Qualitative and quantitative analysis of volatile fatty acids in compost and the growth response of *Agaricus bisporus*. B.Sc. Honors Thesis, La Trobe University, Melbourne, Australia.

Chang, Y. 1967. The fungus of wheat straw compost. II. Biochemical and physiological studies. *Trans. Brit. Mycol. Soc.* 50:649–666.

Chang, Y., and H.J. Hudson. 1967. The fungi of wheat straw compost. I. Ecological studies. *Trans. Brit. Mycol. Soc.* 50:667–677.

Chanyasak, V., T. Yoshida, and H. Kubota. 1980. Chemical components in gel chromatographic fractionation of water extract from sewage sludge compost. *J. Ferment. Technol.* 58:533–539.

Chen, W.A.C., and D.M. Griffin. 1966. Soil physical factors and the ecology of fungi. V. Further studies in relatively dry soils. *Trans. Brit. Mycol. Soc.* 49:419–426.

Ciavatta, C., M. Govi, and P. Sequi. 1993. Characterization of organic matter in compost produced with municipal solid wastes: an Italian approach. *Compost Sci. Utiliz.* 1:75–81.

Cooney, D.G. and E. Emerson. 1964. *Thermophilic Fungi: An Account of Their Biology, Activities and Classification*. W.H. Freeman, San Francisco.

Corominas, E., F. Perestelo, M.L. Perez, and M.A. Falcon. 1987. Microorganisms and environmental factors in composting of agricultural waste of the Canary Islands, in M. DeBertoldi, M.P. Ferranti, P.L. Hermite, and F. Zucconi (Eds.), *Compost: Production, Quality and Use*, Elsevier Science, London, pp. 127–138.

Da Silva, R., D.K. Yim, and Y.K. Park. 1994. Application of thermostable xylanases from *Humicola* sp. for pulp improvement. *J. Ferment. Bioeng.* 77:109–111.

De Bertoldi, M., A. Rutili, B. Citterio, and M. Civillini. 1988. Composting management: a new process control through O_2 feedback. *Waste Manage. Res.* 6:239–259.

De Bertoldi, M., G. Vallini, and A. Pera. 1983. The biology of composting: a review. *Waste Manage. Res.* 1:157–176.

De Bertoldi, M., G. Vallini, A. Pera, and F. Zucconi. 1982. Comparison of three windrow compost systems. *BioCycle* 23:45–50.

Derikx, P.J.L., H.J.M. Op den Camp, C. van der Drift, L.J.L.D. van Griensven, and G.D.Vogels. 1990. Odorous sulfur compounds emitted during production of compost used as a substrate in mushroom cultivation. *Appl. Environ. Microbiol.* 56:176–180.

Dye, M.H. 1964. Self-heating in damp wool. Part 1. The estimations of microbial populations in wool. *N.Z. J. Sci.* 7:87–96.

Dye, M.H., and H.P. Rothbaum. 1964. Self-heating of damp wool. Part 2. Self heating of damp wool under adiabatic conditions. *N.Z. J. Sci.* 7:97–118.

Edwards, C. 1990. Thermophiles, in C. Edwards (Eds.). *Microbiology of Extreme Environments*. McGraw-Hill, New York, pp. 1–33.

Erickson, D. 1952. Temperature/growth relationships of a thermophilic actinomycete, *Micromonospora vulgaris. J. Gen. Microbiol.* 6:286–294.

Federal Register. 1979. 40 CFR Part 257, EPA Part IX. Vol. 44 (179):53438–53468, 13 September 1979.

Fergus, C.L. 1964. Thermophilic and thermotolerant molds and actinomycetes of mushroom compost during peak heating. *Mycologia* 56:286–294.

Fergus, C.L. 1971. Thermophilic and thermotolerant molds and actinomycetes of mushroom compost during peak heating. *Mycologia* 63:426–431.

Fergus, C.L., and J.W. Sinden. 1969. A new thermophilic fungus from mushroom compost: *Thielavia thermophila*, sp. nov. *Can. J. Bot.* 47:1635–1637.

Fermor, T. 1993. Applied aspects of composting and bioconversion of lignocellulosic materials: an overview. *Int. Biodeterior. Biodegrad.* 31:87–106.

Fermor, T.R., P.E. Randle, and J.F. Smith. 1985. Compost as a substrate and its preparation, in P.B. Flegg, D.M. Spencer, and D.A. Wood (Eds.), *The Biology and Technology of the Cultivated Mushroom*, John Wiley & Sons, New York, pp. 81–110.

Fermor, T.R., J.F. Smith, and D.M. Spencer. 1979. The microflora of experimental mushroom composts. *J. Horticult. Sci.* 54:137–147.

Festenstein, G.N., J. Lacey, F.A. Skinner, P.A. Jenkins, and J. Pepys. 1965. Self-heating of hay and grain in Dewar flasks and the development of farmer's lung antigens. *J. Gen. Microbiol.* 41:389–407.

Filip, Z. 1978. Effect of solid particles on the growth and endurance to heat stress of garbage compost microorganisms. *Appl. Microbiol. Biotech.* 6:87–94.

Finstein, M.S., and M.L. Morris. 1975. Microbiology of municipal solid waste composting. *Adv. Appl. Microbiol.* 19:113–151.

Finstein, M.S., K.W. Lin, and G.E. Fischler. 1982. Sludge composting and utilization: review of the literature on the temperature inactivation of pathogens. Report of New Jersey Agricultural Experimental Station Project No. 03543, New Brunswick, NJ.

Finstein, M.S., F.C. Miller, and P.F. Strom. 1986. Waste treatment composting as a controlled system, in H.J. Rehm and G. Reed (Eds.), *Biotechnology*, Vol. 8, VCH Verlagsgesellschaft, Weinheim, FRG, pp. 363–398.

Finstein, M.S., F.C. Miller, J.A. Hogan, and P.F. Strom. 1987a. Analysis of EPA guidance on composting sludge. Part I. Biological heat generation and temperature. *BioCycle* 28:20–25.

Finstein, M.S., F.C. Miller, J.A. Hogan, and P.F. Strom. 1987b. Analysis of EPA guidance on composting sludge. Part II. Biological process control. *BioCycle* 28:42–47.

Finstein, M.S., F.C. Miller, J.A. Hogan, and P.F. Strom. 1987c. Analysis of EPA guidance on composting sludge. Part III. Oxygen, moisture, odor, pathogens. *BioCycle* 28:38–44.

Finstein, M.S., F.C. Miller, J.A. Hogan, and P.F. Strom. 1987d. Analysis of EPA guidance on composting sludge. Part IV. Facility design and operation. *BioCycle* 28:56–61.

Finstein, M.S., F.C. Miller, S.T. Macgregor, and K.M. Psarianos. 1985. The Rutgers strategy for composting: process control and design and control, EPA/600/2-85/059, Accession no. PB85-207538, National Technical Information Service, Springfield, VA.

Finstein, M.S., F.C. Miller, P.F. Strom, S.T. Macgregor, and K.M. Psarianos. 1983. Composting ecosystem management for waste treatment. *Bio/Technol.* 1:347–353.

Forbes, R.H., T.C. Mendenhall, L.O. Young, and J.D. Schwisow. 1994. Combined strategies for odor control. *BioCycle* 35:49–54.

Forsyth, W.G.C., and D.M. Webley. 1948. The microbiology of composting. II. A study of the aerobic thermophilic bacterial flora developing in grass composts. *Proc. Soc. Appl. Bacteriol.* 1948:34–39.

Foster, R.C. 1988. Microenvironments of soil microorganisms. *Biol. Fertility Soils* 6:189–203.

Fujio, Y. and S. Kume. 1991. Isolation and identification of thermophilic bacteria from sewage sludge compost. *J. Ferment. Bioeng.* 72:334–337.

Goldstein, N. 1989. Solid waste composting in the U.S. *BioCycle,* 30:32–37.

Goldstein, N., and R. Steuteville. 1993. Biosolids composting makes healthy progress. *BioCycle* 34:48–57.

Goldstein, N., W.A. Yanko, J.M. Walker, and W. Jakubowski. 1988. Determining pathogen levels in sludge products. *BioCycle* 29:44–47.

Gray, K.R., and K. Sherman. 1969. Accelerated composting of organic wastes. *Birmingham Univ. Chem. Eng.* 20:64–74.

Gray, K.R., K. Sherman, and A.J. Biddlestone. 1971. Review of composting. Part 2. The practical process. *Proc. Biochem.* 6:22–28.

Greenwood, D.J. 1961. The effect of oxygen concentration on the decomposition of organic materials in soil. *Plant Soil* 14:360–376.

Gregory, P.H., M.E. Lacey, G.N. Gestenstein, and F.A. Sinner. 1963. Microbial and biochemical changes during the moulding of hay. *J. Gen. Microbiol.* 33:147–174.

Griffin, D.M. 1981a. Water and microbial stress. *Adv. Microb. Ecol.* 5:91–136.

Griffin, D.M. 1981b. Water potential as a selective factor in the microbiology of soils, in J.F. Parr, W.R. Gardner, and L.F. Elliott (Eds.), *Water Potential Relations in Soil Microbiology*, Soil Science Society of America, Madison, WI, pp. 141–151.

Griffin, D.M. 1985. A comparison of the roles of bacteria and fungi, in E.R. Leadbetter and J.S. Poindexter (Eds.), *Bacteria in Nature*, Vol. I, Plenum Press, New York, pp. 221–255.

Griffin, D.M. and G. Quail. 1968. Movement of bacteria in moist, particulate systems. *Aust. J. Biol. Sci.* 21:579–582.

Haapala, R., S. Linko, E. Parkkinen, and P. Suominen. 1994. Production of endo-1,4-ᵦ-glucanase and xylanase by *Trichoderma reesei* immobilized on polyurethane foam. *Biotechnol.Tech.* 8:401–406.

Hamelers, H.V.M. 1993. A theoretical model of composting kinetics, in H.A.J. Hoitink and H.M. Keener (Eds.), *Science and Engineering of Composting*. Renaissance Publications, Worthington, OH, pp. 36–58.

Harper, E.R., Miller, F.C., and Macauley, B.J. 1992. Physical management and interpretation of an environmentally controlled composting ecosystem. *Aust. J. Exp. Agric.* 32:657–667.

Harris, R.F. 1981. Effect of water potential on microbial growth and activity, in J.F. Parr, W.R. Gardner, and L.F. Elliott (Eds.), *Water Potential Relations in Soil Microbiology*, Soil Science Society of America, Madison, WI, pp. 23–95.

Haug, R.T. 1978. Composting wet organic sludges — a problem of moisture control, Proc. National Conference on Design of Municipal Sludge Composting Facilities, Chicago, IL, pp. 27–38, Information Transfer, Rockville, MD.

Haug, R.T. 1979. Engineering principles of sludge composting. *J. Water Pollut. Control Fed.* 51:2189–2206.

Haug, R.T. 1993. *The Practical Handbook of Compost Engineering*, Lewis Publishers, Chelsea, MI.

Hayes, W.A. 1968. Microbiological and biochemical changes in composting straw/horse manure mixtures. *Mushroom Sci.* 7:173–186.

Hedger, N.J. and H.J. Hudson. 1970. *Thielavia thermophila* and *Sporotrichum thermophila*. *Trans. British Mycol. Soc.* 54:497–500.

Henssen, A. 1957. Uber die bedentung der thrmophilen mickroorganismen fur der zersetzung des stallmistes. *Arch. Fur Mikrobiol.* 27:63–81.

Hogan, J.A., F.C. Miller, and M.S. Finstein. 1989. Physical modeling of the composting ecosystem. *Appl. Environ. Microbiol.* 55:1082–1092.

Hogan, J.A., G.R. Toffoli, F.C. Miller, J.V. Hunter, and M.S. Finstein. 1988. Composting physical model demonstration: mass balance of hydrocarbons and PCBs, pp. 742–758. In Y.C. Wu (Ed.), *International Conference on Physiological and Biological Detoxification of Hazardous Wastes*. Technomic Publishing, Lancaster, PA, pp. 742–758.

Hoitink, H.A.J. 1980. Composted bark, a lightweight growth medium with fungicidal properties. *Plant Dis.* 64:142–147.

Iiyama, K., B.A. Stone, and B.J. Macauley. 1994. Compositional changes in compost during composting and growth of *Agaricus bisporus*. *Appl. Environ. Microbiol.* 60:1538–1546.

Ikasari, L. and D.A. Mitchell. 1994. Protease production by *Rhizopus oligosporus* in solid state fermentation. *World J. Microbiol. Biotechnol.* 10:320–324.

Jeris, J.S., and W.R. Regan. 1973. Controlling environmental parameters for optimum composting. *Compost Sci.* 14:16–22.

Joshua, R.S., B.J. Macauley, and C.R. Hudson. 1994. Recycling grease trap sludges. *BioCycle* 35:46–47.

Kane, B.E., and J.T. Mullins. 1973. Thermophilic fungi in a municipal waste composting system. *Mycologia* 65:1087–1100.

Kerem, Z., D. Friesem, and Y. Hadar. 1992. Lignocellulose degradation during solid state fermentation: *Pleurotus ostreatus* vs. *Phanerochaete chrysosporium*. *Appl. Environ. Microbiol.* 58:1121–1127.

Kissel, J.C., C.L. Henry, and R.B. Harrison. 1992. Potential emissions of volatile and odorous organic compounds from municipal solid waste composting facilities. *Biomass Bioenergy* 3:181–194.

Kleman-Leyer, K.M. and T.K. Kirk. 1994. Three native cellulose depolymerizing endoglucanases from solid substrate cultures of the brown rot fungi *Meruliporia* (Serpula) *incrassata*. *Appl. Environ. Microbiol.* 60:2839–2845.

Koster, I.W. 1986. Characteristics of the pH-influenced adaptation of methanogenic sludge to ammonia toxicity. *J. Chem. Technol. Biotechnol.* 36:445–455.

Kuter, G.A., H.A.J. Hoitink, and L.A. Rossman. 1985. Effects of aeration and temperature on composting of municipal sewage sludge in a full scale vessel system. *J. Water Pollut. Control Fed.* 57:309–315.

Lacey, J. 1973. Actinomycetes in soils, composts and fodders, in G. Sykes and F.A. Skinner (Eds.), *Actinomycetales: Characteristics and Practical Importance,* Academic Press, London, pp. 231–251.

Lambert, E.B. and A.C. Davis . 1934. Distribution of oxygen and carbon dioxide in mushroom compost heaps as affecting microbial thermogenesis, acidity and moisture therein. *J. Agric. Res.* 48:587–601.

Letey, Jr., J., L.H. Stolzy, and W.D. Kemper. 1967. Soil aeration in irrigation of agricultural lands, in R.M. Hagan, H.R. Haise, and T.W. Edminster (Eds.), *Agronomy 11: Irrigation of Agricultural Lands,* Academic Press, New York, pp. 941–949.

Macauley, B.J., B. Stone, K. Iiyama, E.R. Harper, and F.C. Miller. 1993. Composting research runs "hot" and "cold" at La Trobe University. *Compost Sci. Utiliz.* 1:6–12.

Macgregor, S.T., F.C. Miller, K.M. Psarianos, and M.S. Finstein. 1981. Composting process control based on interaction between microbial heat output and temperature. *Appl. Environ. Microbiol.* 41:1321–1330.

Makawi, A.A.M. 1980. The effect of thermophilic actinomycetes isolated from compost and animal manure on some strains of *Salmonella* and *Shigella. Zeit. Bakteriol.* 135:12–21.

McCauley, R.F., and B.J. Shell. 1956. Laboratory and operational experiences in composting, *Proc. Purdue Ind. Waste Conf.* 11:436–453.

McKinley, V.L., and J.R. Vestal. 1984. Biokinetic analysis of adaptation and succession: microbial activity in composting municipal sewage sludge. *Appl. Environ. Microbiol.* 47:933–939.

Miller, F.C. 1984. Thermodynamic and matric water potential analysis in field and laboratory scale composting ecosystems. Ph.D. Dissertation, Rutgers University, University Microfilms, Ann Arbor, MI.

Miller, F.C. 1989. Matric water potential as an ecological determinant in compost, a substrate dense system. *Microb. Ecol.* 18:59–71.

Miller, F.C. 1990. *Biological Treatment of Sewage Sludge or Similar Wastes.* International application published under the Patent Cooperation Treaty. International Publication Number WO 90/09964, World Intellectual Property Organization, International Bureau.

Miller, F.C. 1992. Minimizing odor generation, in H.A.J. Hoitink and H. M. Keener (Eds.), *Science and Engineering of Composting: Design, Environmental, Microbiological and Utilization Aspects.* Renaissance Publications, Worthington, OH, pp. 219–241.

Miller, F.C. 1994. Mushroom composting as a managed ecosystem. *Mushroom Res.* 3:5–13.

Miller, F.C., and M.S. Finstein. 1985. Materials balance in the composting of sewage sludge as affected by process control strategy. *J. Water Pollut. Control Fed.* 57:122–127.

Miller, F.C., and B.J. Macauley. 1988. Odours arising from mushroom composting: a review. *Aust. J. Exp. Agric.* 28:553–560.

Miller, F.C., and B.J. Macauley. 1989. Substrate usage and odours in mushroom composting. *Aust. J. Exp. Agric.* 29:119–124.

Miller, F.C., E.H. Harper, B.J. Macauley, and A. Gulliver. 1990. Composting based on moderately thermophilic and aerobic conditions for the production of mushroom compost. *Aust. J. Exp. Agric.* 30:287–296.

Miller, F.C., E.R. Harper, and B.J. Macauley. 1989. Field examination of temperature and oxygen relationships in mushroom composting stacks — consideration of stack oxygenation based on utilisation and supply. *Aust. J. Exp. Agric.* 29:741–749.

Miller, F.C., B.J. Macauley, and E.R. Harper. 1991. Investigation of various gases, pH and redox potential in mushroom composting Phase I stacks. *Aust. J. Exp. Agric.* 30:415–425.

Miller, F.C., S.T. Macgregor, M.S. Finstein, and J. Cirello. 1980. Biological drying of sewage sludge — a new composting process. Proc. ASCE, Environmental Engineering Division Specialty Conference, American Society of Civil Engineers, New York, 1980, pp. 40–49.

Miller, F.C., S.T. Macgregor, K.M. Psarianos, J. Cirello, and M.S. Finstein. 1982. Direction of ventilation in composting wastewater sludge. *J. Water. Pollut. Control Fed.* 54:111–113.

Millner, P.D., P.B. Marsh, R.B. Snowden, and J.F. Parr. 1977. Occurrence of *Aspergillus fumigatus* during composting of sewage sludge. *Appl. Environ. Microbiol.* 34:765–772.

Mortland, M.M., and A.R. Wolcott. 1965. Sorption of inorganic nitrogen compounds by soil materials, in W.V. Bartholomew and F.E. Clark (Eds.), *Soil Nitrogen.* American Society of Agronomy, Madison, WI, pp. 151–197.

Myers, R.G., and S.J. Thien. 1988. Organic matter solubility and soil reaction in an ammonium and phosphorous application zone. *Soil Sci. Soc. Am. J.* 52:516–522.

Nakasaki, K., N. Aoki, and H. Kubota. 1994. Accelerated composting of grass clippings by controlling moisture level. *Waste Manage. Res.* 12:12–20.

Nakasaki, K., Y. Nakano, T. Akiyama, M. Shoda, and H. Kubota. 1987. Oxygen diffusion and microbial activity in the composting of dehydrated sewage sludge cakes. *J. Ferment. Technol.* 65:43–48.

Nakasaki, K., H. Yaguchi, Y. Sasaki, and H. Kubota. 1993. Effects of pH control on composting of garbage. *Waste Manage. Res.* 11:117–125.

Nell, J.H., and S.G. Wiechers. 1978. High temperature composting. *Water South Africa* 4:203–212.

Niese, G. 1959. Mikrobiologische untersuchungne zur frage der selbsterhitzung organischer stoffe. *Arch. Fur Microbiol.* 34:285–318.

Nommik, H. 1965. Ammonia fixation and other reactions involving a nonenzymatic immobilization of mineral nitrogen on soil, in W.V. Bartholomew and F.E. Clark (Eds.), *Soil Nitrogen.* American Society of Agronomy, Madison, WI, pp. 198–258.

Nommik, H. 1970. Non-exchangeable binding of ammonium and amino nitrogen by Norway spruce raw humus. *Plant & Soil* 33:581–595.

Norman, A.G. 1930. The biological decomposition of plant materials. Part III. Physiological studies on some cellulose decomposing fungi. *Ann. Appl. Biol.* 17:575–613.

Norman, A.G., L.A. Richards, and R.E. Carlyle. 1941. Microbial thermogenesis in the decomposition of plant materials. Part I. An adiabatic fermentation apparatus. *J. Bacteriol.* 41:689–697.

Ohno, A., T. Ano, and M. Shoda. 1992. Production of antifungal antibiotic, Iturin, in a solid state fermentation by *Bacillus subtilis* NB22 using wheat bran as a substrate. *Biotechnol. Lett.* 14:817–822.

Okafor, N. 1966. Thermophilic microorganisms from rotting maize. *Nature* 210:220–221.

Op den Camp, H.J.M. 1989. Aeroob *vs.* anaeroob: de vorming van mathaan tijdens composteren, *De Champignonculture* 31:513–519.

Overstijns, A. 1994. Indoor composting system, in Proc. Second AMGA/ISMS International Workshop — Seminar on Agaricus Compost, Australian Mushroom Growers Association, Aug. 29–Sept. 3, 1993, Windsor, NSW, Australia.

Papendick, R.I., and G.S. Campbell. 1981. Theory and measurement of water potential, in J.F. Parr, W.R. Gardner, and L.F. Elliott (Eds.), *Water Potential Relations in Soil Microbiology.* Soil Science Society of America, Madison, WI, pp. 1–22.

Parr, J.F., E. Epstein, and G.B. Willson. 1978. Composting sewage sludge for land application. *Agric. Environ.* 4:123–137.

Poincelot, R.P. 1975. The biochemistry and methodology of composting, Bulletin 754, The Connecticut Agricultural Experiment Station, New Haven, CT.

Richard, T. L. 1992. Municipal solid waste composting: physical and biological processing. *Biomass Bioenergy* 3:163–180.

Richard, T.L. and P.B. Woodbury. 1992. The impact of separation on heavy metal contaminants in municipal solid waste composts. *Biomass Bioenergy* 3:195–211.

Robinson, J.J. and E.I. Stentiford. 1993. Improving the aerated static pile composting method by the incorporation of moisture control. *Compost Sci. Util.* 1:52–68.

Ross, R.C, and P.J. Harris. 1982. Some factors involved in Phase II of mushroom compost preparation. *Sci. Hortic.* 17:223–229.

Ross, R.C., and P.J. Harris. 1983a. An investigation into the selective nature of mushroom compost. *Sci. Hortic.* 19:55–64.

Ross, R.C. and P.J. Harris. 1983b. The significance of thermophilic fungi in mushroom compost preparation. *Sci. Hortic.* 20:61–70.

Rothbaum, H.P. 1961. Heat output of thermophiles occurring on wool. *J. Bacteriol.* 81:165–171.

Rothbaum, H.P. 1963. Spontaneous combustion of hay. *Appl. Chem.* 13:291–302.

Rothbaum, H.P., and M.H. Dye. 1964. Self-heating of damp wool. Part 3. Self-heating of damp wool under isothermal conditions. *N.Z. J. Sci.* 7:119–146.

Samson, R.A., M.J. Crisman, and M.R. Tansey. 1977. Observations on the thermophilic ascomycete *Thielavia terrestris. Trans. Brit. Mycol. Soc.* 69:417–423.

Schulze, K.L. 1958. Rate of oxygen consumption and respiratory quotients during the aerobic decomposition of a synthetic garbage. *Proc. Purdue Ind. Waste Conf.* 13:541–554.

Schulze, K.L. 1961. Relationship between moisture content and activity of finished compost. *Compost Sci.* 2:32–34.

Schulze, K.L. 1962. Continuous thermophilic composting. *Appl. Microbiol.* 10:108–122.

Sharon, N. 1965. Distribution of amino sugars in microorganisms, plants, and invertebrates, in E.A. Balaz and R.W. Jeanloz (Eds.), *The Amino Sugars — The Chemistry and Biology of Compounds Containing Amino Sugars,* Vol. IIA, Academic Press, New York, pp. 1–45.

Shell, G.L., and J.L. Boyd. 1969. *Composting Dewatered Sewage Sludge.* Report SW-12c, USDHEW/Public Health Service, Environmental Health Services Environmental Control Administration, Bureau of Solid Waste Management, Washington, D.C.

Stanek, M. 1971. Microorganisms inhabiting mushroom compost during fermentation. *Mushroom Sci.* 8:797–811.

Steuteville, R. 1994. The state of garbage in America. *BioCycle* 35:46–52.

Stevenson, F.J. 1982. *Humus Chemistry: Genesis, Composition, Reactions*, Wiley Interscience, New York.

Straatsma, G., R.A. Samson, T.W. Olijnsma, H.J.M. Op den Camp, J.P.G. Gerrits, and L.J.L.D. Van Griensven. 1994. Ecology of thermophilic fungi in mushroom compost, with emphasis on *Scytalidium thermophilum* and growth stimulation of *Agaricus bisporus* mycelium. *Appl. Environ. Microbiol.* 60:454–458.

Strom, P.F. 1985a. Effect of temperature on bacterial diversity in thermophilic solid waste composting. *Appl. Environ. Microbiol.* 50:899–905.

Strom, P.F. 1985b. Identification of thermophilic bacteria in solid-waste composting, *Appl. Environ. Microbiol.* 50:906–913.

Stutzenberger, F.J. 1971. Cellulase production by *Thermomonospora curvata* isolated from municipal solid waste compost. *Appl. Microbiol.* 22:147–152.

Sudo, S., T. Ishikawa, K. Sato, and T. Oba. 1994. Comparison of acid stable alphaamylase production by *Aspergillus kawachii* in solid state and submerged cultures. *J. Ferment. Bioeng.* 77:483–489.

Suler, D.J., and M.S. Finstein. 1977. Effect of temperature, aeration, and moisture on CO_2 formation in bench-scale, continuously thermophilic composting of solid wastes. *Appl. Environ. Microbiol.* 32:345–350.

Sundaram, T.K. 1986. Physiology and growth of thermophilic bacteria, in T.D. Brock (Ed.), *Thermophiles — General, Molecular, and Applied Microbiology*, John Wiley & Sons, New York, pp. 75–106.

Tanimoto, Y. and F. Bak. 1994. Anaerobic degradation of methylmercaptan and dimethyl sulfide by newly isolated thermophilic sulfate reducing bacteria. *Appl. Environ. Microbiol.* 60:2450–2455.

Tansey, M.R. 1981. Isolation of thermophilic fungi from self-heated, industrial wood chip piles. *Mycologia* 58:537–547.

Tate, III, R.L. 1987. *Soil Organic Matter: Biological and Ecological Effects*, Wiley Interscience, New York.

Tchobanoglous, G., H. Theisen, R. Eliassen. 1977. *Solid Wastes: Engineering Principles and Management Issues*, McGraw-Hill, New York.

Tendler, M.D. and P.R. Burkholder. 1961. Studies on the thermophilic actinomycetes. I. Methods of cultivation. *Appl. Microbiol.* 9:394–399.

U.S. Congress, Office of Technology Assessment. 1989. *Facing America's Trash: What Next for Municipal Solid Waste*. OTA-O-424. U.S. Government Printing Office, Washington, D.C.

Waksman, S.A., and T.C. Cordon. 1939. Thermophilic decomposition of plant residues in compost by pure and mixed cultures of micro-organisms. *Soil Sci.* 47:217–225.

Waksman, S.A., and I.J. Hutchings. 1937. Associative and antagonistic effects of micro-organisms. III. Associative and antagonistic relationships in the decomposition of plant residues. *Soil Sci.* 43:77–92.

Waksman, S.A., T.C. Cordon, and N. Hulpoi. 1939a. Influence of temperature on the microbiological population and decomposition processes in composts of stable manure. *Soil Sci.* 47:83–114.

Waksman, S.A., W.W. Umbreit, and T.C. Cordon. 1939b. Thermophilic actinomycetes and fungi in soils and in composts. *Soil Sci.* 47:37–61.

Walker, I.K., and W.J. Harrison. 1960. The self heating of wet wool. *N.Z. J. Agric. Res.* 3:861–895.

Wiacek-Zychlinska, A., J. Czakaj, and R. Sawicka-Zukowska. 1994. Xylanase production by fungal strains in solid-state fermentations. *Bioresource Technol.* 49:13–16.

Wiley, J.S. 1955. Studies of high-rate composting of garbage and refuse. *Proc. Purdue Ind. Waste Conf.* 11:436–453.

Wiley, J.S. 1956. Progress report on high-rate composting studies. *Proc. Purdue Ind. Waste Conf.* 11:596–603.

Wiley, J.S. 1957. Liquid content of garbage and refuse. Proc. ASCE, Sanitary Engineering Division, American Society of Chemical Engineers, Washington, D.C., p. 1411.

Willson, G.B., J.F. Parr, E. Epstein, P.B. Marsh, R.L Chaney, W.D. Colcicco, W.D. Burge, L.J. Sikora, C.F. Tester, and S. Hornic. 1980. *Manual for Composting Sewage Sludge by the Beltsville Aerated Pile Method*. U.S. Environmental Protection Agency, U.S. Department of Agriculture, U.S. Government Printing Office, Washington, D.C.

Wimpenny, J.W.T., P.J. Coombs, and R.W. Lovitt. 1984. Growth and interactions of microorganism in spatially heterogeneous ecosystems, in *Current Perspectives in Microbial Ecology — Proc. Third Int. Symp. Microbial Ecology,* American Society for Microbiology, Washington, D.C., pp. 291–299.

Witter, E. and J.M. Lopez-Real. 1987. The potential of sewage sludge and composting in a nitrogen recycling strategy for agriculture. *Biol. Agric. Horticult.* 5:1–23.

Wood, D.A. 1984. Microbial processes in mushroom cultivation: large scale solid substrate fermentation. *J. Chem. Technol. Biotechnol.* 34 B:232–240.

Yang, S.S., H.D. Jang, C.M. Liew, and J.C. du Preez. 1993. Protein enrichment of sweet potato residue by solid state cultivation with mono- and co-cultures of amylolytic fungi. *World J. Microbiol. Biotechnol.* 9:258–264.

Zwietering, M.H., J.T. de Koos, B.E. Hasenack, J.C. de Wit, and K. van 'T Riet. 1991. Modeling of bacterial growth as a function of temperature. *Appl. Environ. Microbiol.* 57:1094–1101.

5

Microbial Pathogens in Municipal Solid Waste

Charles P. Gerba

CONTENTS

5.1 Introduction 156

5.2 Pathogens of Concern 156
    5.2.1 Bacteria..156
    5.2.2 Enteric Bacteria..157
    5.2.3 Enteric Protozoa..158
    5.2.4 Enteric Viruses 159
    5.2.5 Fungi...160

5.3 Sources of Pathogens in Municipal Solid Waste 161
    5.3.1 Sludge Biosolids 161
    5.3.2 Domestic Pet Waste 162
    5.3.3 Disposable Diapers 163

5.4 Occurrence and Survival of Enteric Microorganisms in Landfills 164

5.5 Occurrence and Survival of Enteric Organisms in Leachate 166

5.6 Fate of Pathogens During Composting of Municipal Solid Waste....................168
    5.6.1 Enteric Pathogens...168
    5.6.2 Fungal Pathogens 169

5.7 Conclusions 170

References 171

0-8493-8361-7/96/$0.00+$.50
© 1996 by CRC Press, Inc.

5.1 INTRODUCTION

Components of municipal solid waste (MSW) entering landfills may contain a variety of pathogenic bacteria, viruses, and parasites. Pathogens of concern usually originate from humans or animals who are infected and they can occur on the skin, saliva, urine, and feces. The major concern is with enteric pathogens because of their potential for long-term survival in the environment. Enteric pathogens are transmitted by the fecal-oral route and may be transmitted by either contact with fecally contaminated material or ingestion of contaminated food or water. Enteric pathogens may be present in food wastes, disposable diapers, pet wastes, sewage sludge, and septic tank wastes. Information on the actual concentrations and fate of pathogens in MSW is limited. Most of the research on this topic was conducted in the 1970s in an attempt to determine the potential for groundwater contamination by enteric viruses. Unfortunately, many new groups of pathogens have been discovered which may exhibit greater survival in the environment or potentially could grow in MSW. Although, overall, the risk from pathogens in solid waste appears to be minimal if it is properly handled and disposed, pathogens always should be reevaluated as new information becomes available. Table 1 summarizes the major human enteric pathogens of concern, including protozoa, bacteria, and viruses. In this chapter, information on the sources and concentration of these enteric pathogens is provided, as well as their fate and survival in landfills, landfill leachate, and compost.

5.2 PATHOGENS OF CONCERN

5.2.1 Bacteria

Enteric bacteria of concern include *Salmonella, Shigella, Campylobacter, Yersinia, Vibrio cholera*, and pathogenic strains of *Escherichia coli*. Also of interest are the fecal indicator bacteria such as coliforms, fecal coliforms, and fecal streptococci. Isolation of enteric pathogens from the environment is difficult, and indicator bacteria provide an indirect means of assessing their potential sources and fate in the environment. The major illness caused by enteric bacteria is gastroenteritis. Gastroenteritis caused by the enteric bacteria is usually more severe than gastroenteritis caused by the enteric viruses and protozoan parasites, and it results in greater mortality, especially in the very young and elderly. *Shigella* is the only enteric pathogen in this group that originates only from man. The other enteric bacterial pathogens may have significant animal reservoirs. *Salmonella* and *Campylobacter* are major agents of foodborne disease in the U.S. Studies have shown that 30% or more of the poultry sold on the market may be contaminated with these bacteria.

TABLE 1   Major Human Enteric Pathogens

Group	Pathogen	Disease
Protozoa	*Giardia lamblia*	Diarrhea
	Cryptosporidium	Diarrhea
Bacteria	*Salmonella*	Typhoid, diarrhea
	Shigella	Diarrhea
	Campylobacter	Diarrhea
	Vibrio cholerae	Diarrhea
	Yersinia enterocolitica	Diarrhea
	Escherichia coli (certain strains)	Diarrhea
Viruses	Enteroviruses (polio, echo, coxsackie)	Meningitis, paralysis, rash, fever, myocarditis, respiratory disease, diarrhea, diabetes
	Hepatitis A and E	Hepatitis
	Norwalk virus	Diarrhea
	Rotavirus	Diarrhea
	Astrovirus	Diarrhea
	Calcivirus	Diarrhea
	Adenovirus	Diarrhea, eye infections, respiratory disease
	Reovirus	Respiratory, enteric

5.2.2 Enteric Bacteria

Salmonella is a very large group of organisms consisting of more than 2000 known serotypes. All serotypes are pathogenic to humans and are capable of causing mild gastroenteritis to severe illness and death. *Salmonella* are capable of infecting a large variety of both cold and warm blooded animals. Typhoid fever, caused by *S. typhi*, and paratyphoid fever, caused by *S. typhoid*, are both enteric fevers that occur only in humans and primates. In the U.S., salmonellosis is primarily due to foodborne transmission since the bacteria infect beef and poultry and are capable of growing in foods.

Shigella sp. only infects human beings and causes gastroenteritis and fever. It does not appear to survive long in the environment, but outbreaks from drinking untreated water and swimming in sewage-contaminated surface waters continue to occur in the U.S. *Campylobacter* and *Yersinia* sp. occur in fecally contaminated water and food, and they are largely believed to originate from animal feces. *Campylobacter* commonly infects poultry, and contaminated foods are regularly implicated as sources of outbreaks. *Campylobacter* continues to be associated with the consumption of untreated drinking water in the U.S. *Escherichia coli* is found in the gastrointestinal tract of all warm blooded animals and is usually considered a harmless organism; however, several strains are capable of causing gastroenteritis. These are referred to as *enterotoxigenic* (ETEC), *enteropathogenic* (EPEC), *enterohemorrhagic* (EHEC), and *enteroinvasive* strains of *E. coli*. Enterotoxigenic *E. coli* causes gastroenteritis with profuse watery diarrhea accompanied by nausea, abdominal cramps, and vomiting. Outbreaks originating from both contaminated food and water have been caused by this organism.

The routine examination of water for the presence of intestinal pathogens is currently a tedious, difficult, and time-consuming task. To overcome this problem, the indicator concept was developed at the turn of the last century. It was recognized that certain nonpathogenic bacteria occur in feces of all warm blooded animals and that they easily could be isolated and quantified by simple bacteriological methods. Their presence in water could be considered an indication of fecal contamination and the concurrent presence of enteric pathogens. Coliform bacteria wcre found to be often associated with the presence of fecal contamination and could be easily differentiated from other bacteria which occur in the environment.

Coliform bacteria occur normally in the intestines of all warm-blooded animals and are excreted in great numbers in feces. In polluted water, coliform bacteria are found in densities roughly proportional to the degree of fecal pollution. Coliform bacteria generally are more hardy than disease-causing bacteria; therefore, their absence from water is an indication that the water is bacteriologically safe for human consumption. The presence of the coliform group of bacteria is indicative that other kinds of microorganisms capable of causing disease also may be present and that the water is unsafe to drink.

The coliform group includes all aerobic and facultatively anaerobic, gram-negative, non-spore-forming, rod-shaped bacteria which ferment lactose with gas production in prescribed culture media within 48 hours at 35°C. Coliform bacteria include *Escherichia, Citrobacter, Enterobacter,* and *Klebsiella* species. Fecal coliform bacteria are more specific indicators of fecal contamination. They are differentiated in the laboratory by their ability to ferment lactose, with the production of acid and gas at 44.5°C within 24 hours. Although this test does determine coliforms of fecal origin, it does not distinguish between human and animal contamination.

5.2.3 Enteric Protozoa

The protozoa *Cryptosporidium* and *Giardia* are major causes of waterborne illness in the U.S. (Rose, 1990). *Giardia* is the most common cause of waterborne illness in the U.S. when an agent can be identified. It is believed that at least 60% of the giardiasis in the U.S. results from drinking or contact with contaminated water (Bennett et al., 1987). *Cryptosporidium* has caused several large waterborne outbreaks in the U.S. and the U.K. Both *Giardia* and *Cryptosporidium* may be transmitted by contact with inanimate objects, ingestion or handling of food, or swimming. Both *Giardia* and *Cryptosporidium* have significant animal reservoirs. The human pathogen *Giardia lamblia* infects both beavers and muskrats. *Cryptosporidium parvum* infects many domesticated animals such as cows and horses. Domestic dogs and cats also may be sources of these organisms. (Lewis, 1988; Ungar, 1990). *Giardia* and *Cryptosporidium* cause gastroenteritis, which usually is self-limiting in normal healthy individuals. Giardiasis may last for 2 to 6 months if untreated, while cryptosporidiosis usually lasts 10 days or less; however, *Cryptosporidium* infections often can result in death in immunocomprised individuals such as AIDS patients.

Giardia can form a resistant cyst, and *Cryptosporidium* produces a resistant oocyst. These forms are very stable in the environment and may survive for many months at temperatures below 10°C (DeRegnier et al., 1989). They are rapidly inactivated, however, at temperatures above 40°C (Anderson, 1985). They also are inactivated at freezing temperatures and during drying on surfaces. Both are more resistant than enteric bacteria and viruses to disinfectants used in water and wastewater treatment (i.e., chlorine, ozone). The oocysts of *Cryptosporidium* are the most resistant enteric pathogen known to chemical disinfection.

5.2.4 Enteric Viruses

More than 140 different types of viruses are known which infect the intestinal tract of humans and are excreted in their feces. Viruses which infect and multiply in the intestines are referred to as *enteric viruses*. Some enteric viruses are capable of replicating in other organs, such as the liver, nerve tissue, the eye, and the heart. For example, hepatitis A virus infects the liver, causing hepatitis. Enteric viruses are very host specific, and, generally, human enteric viruses cause disease only in humans and sometimes primates. During infection, large numbers of viruses may be excreted in the feces, 10^8 to 10^{12} per gram (Gerba and Rose, 1993). All of these viruses are capable of being transmitted by food or fomites.

Enteroviruses were the first enteric viruses ever isolated from sewage and water, and they have been the most extensively studied. The more common enteroviruses include the polioviruses (3 types), coxsackieviruses (30 types), and the echoviruses (34 types). They are capable of causing a wide range of serious illness, although most infections are mild (Gerba and Rose, 1993). Usually, only 50% of the people infected actually develop clinical illness. However, coxsackieviruses can cause a number of life-threatening illnesses including heart disease, meningitis, paralysis, and insulin-dependent diabetes.

Infectious viral hepatitis is caused by hepatitis A virus (HAV) and hepatitis E virus (HEV). These types of viral hepatitis are spread by fecally contaminated water and food; whereas, other types of viral hepatitis (i.e., hepatitis B virus, or HBV) are spread by exposure to contaminated blood. Hepatitis A and E virus are very common infections in the developing world, with as many as 98% of the population having antibodies against HAV. HAV is not only associated with waterborne outbreaks, but is also commonly associated with foodborne outbreaks, especially shellfish (Gerba et al., 1985). HEV has been associated with large waterborne outbreaks in Asia and Africa, but no outbreaks have been documented in developed countries (Ticehurst et al., 1991). HAV is one of the enteric viruses most resistant to inactivation by heat.

The Norwalk virus was first discovered in 1968 after an outbreak of gastroenteritis in Norwalk, OH. The illness caused by the virus is characterized by vomiting and diarrhea which lasts a few days. It is the agent most commonly identified during water- and foodborne outbreaks of viral gastroenteritis in the U.S. (Gerba et al., 1985). The virus has not yet been grown in the laboratory, but appears to be related to the calcivirus group.

Rotaviruses (five types) are the major cause of infantile acute gastroenteritis in children under 2 years of age. They are a major cause of mortality in children and are responsible for millions of childhood deaths per year in Africa, Asia, and Latin America. They also are responsible for outbreaks among adult populations (especially the elderly) and are a major cause of traveler's diarrhea. Several waterborne outbreaks have been associated with rotaviruses (Gerba et al., 1985).

Adenoviruses are capable of causing a variety of illnesses, including diarrhea and eye, throat, and respiratory infections. They are excreted both in respiratory secretions and in the feces. Enteric adenoviruses (types 40 and 41) are the second most common cause of enteric gastroenteritis in children (Cruz et al., 1990).

Enteric viruses appear to be more stable in the environment than any enteric bacteria or protozoa parasite, and they may survive for prolonged periods of time at low temperatures. Moreover, the enteric viruses usually are more resistant to inactivation at high temperatures ($>50°C$) than the other pathogens. Hepatitis A virus and enteric adenoviruses 40 and 41 appear to be the most stable at high temperatures (Enriquez et al., 1995).

5.2.5 Fungi

Fungi are eucaryotic organisms that play a major role in the decomposition of organic matter. A limited number are pathogenic to humans and cause fungal diseases called *mycoses*. Some airborne fungal spores are responsible for allergies in humans. Fungi usually are considered opportunistic pathogens and cause infection only when an individual is exposed to large numbers, is already debilitated by other diseases, or is immunocompromised. Because fungi grow to large numbers in composting material, concern has arisen over the possible growth of opportunistic species in compost. To date, the greatest amount of interest has centered on the genus *Aspergillus* which can cause aspergillosis. *A. fumigatus* causes the most serious infections among the genus *Aspergillus*. *A. fumigatus* is commonly found in composting vegetation, woodchip piles, MSW compost, refuse-sludge compost, and moldy hay (Millner et al., 1977). *A. fumigatus* grows over a range of below 20°C to about 50°C. Spores of this fungus can cause bronchopulmonary hypersensitivity, marked by asthmatic spasm, fever, and malaise (Emmons et al., 1979).

5.3 SOURCES OF PATHOGENS IN MUNICIPAL SOLID WASTE

Waste material disposed in landfills may originate from numerous sources, including household, commercial, construction, and infectious waste and sewage sludge (see Chapter 1). The overall contribution of sewage sludge is small, but may vary significantly with each community, depending on the availability of other means of disposal (Yanko, 1989); however, sewage sludge can be a significant contributor to the pathogen load in MSW.

As described in detail in Chapter 1, MSW primarily consists of paper, fines, and organics such as yard waste, wood, and food. Disposable diapers are estimated to comprise 0.53 to 1.82% of the total percentage in pounds of waste (Rathje and Murphy, 1992). Pahren (1987) states that the greatest concentrations of enteric bacterial indicators (i.e., total coliforms, fecal coliforms, and fecal streptococci) have occurred in paper, garden waste, and food. Fines are also a significant contributor in some years. The contribution of diapers usually was less, but highly variable. Animal feces are probably the major source of fecal bacteria in yard waste and fines. Overall, the most likely sources of enteric pathogens appear to be sewage sludge, pet wastes, food wastes, and disposable diapers. The numbers of fecal indicator bacteria in sewage sludge, hospital wastes, and municipal wastes are shown in Table 2. High concentrations of those indicators occur in wastes that may end up in landfills. Several species of *Salmonella* have been isolated from raw refuse (Pahren, 1987).

TABLE 2   Concentrations of Indicator Organisms in Various Sources of Waste

	Organisms per gram (dry weight)		
	Sewage Sludge	**Hospital Wastes**	**Municipal Wastes**
Total coliforms	2.8×10^9	9.0×10^8	7.7×10^8
Fecal coliforms	2.4×10^8	9.0×10^8	4.7×10^8
Fecal streptococci	3.3×10^7	8.6×10^8	2.5×10^9

Source: Compiled from Pahren, H.R. 1987. *CRC Crit. Rev. Environ. Control* 17:187-228; Donnelly, J.A., and P.V. Scarpino. 1983. EPA-600/2-84-119. U.S. Environmental Protection Agency, Cincinnati, OH. With permission.

5.3.1 Sludge Biosolids

Co-disposal of sewage sludge or biosolids with MSW has been practiced often in the U.S. (USEPA, 1986). The concentration of microorganisms present in sewage sludge depends on the type of sludge and the treatment it receives before disposal (Sagik et al., 1980). Raw and untreated sludges will contain significantly higher numbers of enteric pathogens than secondary and treated or dried sludges. It is common practice in the U.S. to stabilize sewage sludges by anaerobic or aerobic digestion before disposal, which significantly reduces the level of pathogenic microorganisms. Today, almost all sewage sludge is treated before disposal. To reduce the volume of sludge, it is often vacuum-dried after treatment and prior to disposal, a process that also may reduce the

TABLE 3   Concentrations of Pathogens in
Secondary Sewage Sludge

Organism	Density (number per gram dry weight)
Bacteria	
Total coliforms	7×10^8
Fecal coliforms	8×10^8
Fecal streptococci	2×10^2
Salmonella	9×10^2
Viruses	
Enteroviruses	$10^2 - 10^4$
Parasites	
Ascaris sp.	1×10^3
Trichuris vulpis	1×10^1
Toxocara sp.	3×10^2
Giardia	$1 \times 10^2 - 2 \times 10^3$

Source: Compiled from Fradkin, L. et al. 1989. *J. Environ. Health* 51:148-152; Soares, A.C. et al. 1994. *J. Environ. Sci. Health* A29:1987-1997. With permission.

level of pathogenic organisms. The concentrations of various pathogens found in secondary (activated sludge) municipal sludge are shown in Table 3.

Levels of *Salmonella* in secondary sludge have been reported to be approximately 9×10^2 colony forming units (CFU)/g dry weight. Total coliforms and fecal coliforms occur in much greater concentrations of 10^6 to 10^8 CFU/g than *Salmonella*. Dudley et al. (1980) also reported similar levels of indicator bacteria in sludges undergoing land application. Levels of viruses in sludges were reported by Sorber and Moore (1986); the numbers of enteroviruses ranged from 10^2 to 10^4 plaque forming units (PFU)/g total suspended solids in a study by Fradkin et al. (1989). Safferman et al. (1988) found indigenous enteroviruses in primary and activated sludges. The mean number of viruses in activated sludge was 164 PFU per 100 ml.

Information on the presence of protozoan parasites is limited because of the lack of efficient methods for their detection in biosolids. The concentration of *Giardia* in anaerobically digested sludge has been reported to range from 1.00×10^5 to 4.1×10^6 cysts per kg (Soares et al., 1994).

5.3.2 Domestic Pet Waste

Pet waste, including dog feces in yard waste and cat litter, are household items likely to be discarded in landfills. Pet waste has not been documented as a waste category, presumably due to the inability to distinguish this material from yard waste or soil; however, Matsuto and Ham (1990) noted the presence of cat litter in soil from a sample of refuse from Madison, WI. Newspaper and plastic bags are often associated with domestic fecal matter and may become contaminated with microorganisms (Pahren, 1987).

More than 150 diseases may be transmitted between animals and man. Enteroviruses such as echovirus and coxsackievirus are believed to infect both domestic animals and man (Clapper, 1970). This is thought to be due to the close association of animals with humans.

Concentrations of enteric bacteria isolated from domestic animal stools range from 10^6 to 10^8 per gram stool wet weight. The ratio of fecal coliforms to fecal streptococci in MSW as reported by Donnelly and Scarpino (1983) was 0.2, which is indicative of a nonhuman source. Pahren (1987) states that this is expected since most of the fecal material in municipal waste is of animal origin.

Bacterial levels in dog feces, as determined by Grew et al. (1970), were 2.3×10^7 CFU/g wet weight for fecal coliforms and 9.8×10^8 CFU/g wet weight for fecal streptococci. Levels of fecal coliforms and fecal streptococci in cat feces were 7.9×10^6 and 2.5×10^7 CFU/g wet weight, respectively.

Domestic dogs and cats can become infected with *Giardia* (Lewis, 1988; Swabby et al., 1988) and *Cryptosporidium* (Ungar, 1990). The prevalence of *Giardia* cysts in dog feces has been reported to range from 1 to 68% (Lewis, 1988). The wide range of reported incidents may reflect differences in sampling methods (Douglas et al., 1988) or differences in the incidence in different communities.

The presence of enteric viruses in dogs has been documented by Lundgren et al. (1968) who isolated coxsackievirus B1, B3, and B5 from beagles. Clapper (1970), who isolated echovirus 6 and coxsackievirus B1, B3, and B5 from dogs, concluded that the dog isolates were identical to human strains and that human enteroviruses may be carried in the intestinal tract of animals. Grew et al. (1970) found poliovirus type 1 to be the most frequently isolated virus from domestic animals; however, coxsackievirus A20 and A9 also were isolated.

5.3.3 Disposable Diapers

It has been estimated that 2 million tons of soiled diapers are deposited annually in landfills and that diapers make up 1% of the dry weight of solid waste disposed of in landfills in the U.S. (Pahren, 1987). Diapers soiled with fecal material, such as the other materials listed above, are a potential source of infectious microorganisms. Peterson (1974) found that 33% of the diapers examined were soiled with fecal material and calculated that there were approximately 0.2 g infant feces per pound solid waste (wet weight) or one pound infant feces per ton (based on the fact that diapers make up 1% of solid waste and that one third are fecally contaminated).

Peterson (1972) found poliovirus type 3 (146 PFU/g feces) in one out of ten fecally soiled diapers sampled. In a later study, Peterson (1974) found that 9 of 89 fecally soiled diapers were positive for poliovirus type 3 and echovirus. Other enteric viruses, such as rotavirus, adenovirus, and astroviruses, are likely to occur in soiled diapers, since they are common causes of childhood gastro-enteritis. Thus, the total number of enteric viruses in infant stools may be significantly higher than that reported by Peterson (1974).

No attempt has been made to quantitate the occurrence of enteric protozoa in infant feces, although enteric protozoa commonly cause infection in this age group. The overall incidence of giardiasis in the U.S. is believed to be 3.8% (Howell and Waldron, 1978); however, the incidence in children is likely to be higher. The percentage of children under 2 years of age who are excreting *Giardia* has been reported to range from 8 to 26% in daycare centers and 1 to 3% for children not in daycare centers (Black et al., 1977; Sealy and Shuman, 1983; Picking et al., 1984). The incidence of *Cryptosporidium* in children averages 4% of those with gastroenteritis, but the incidence in asymptomatic children in daycare centers has been reported to be as high as 3 to 27% (Ungar, 1990). The concentration of *Giardia* cysts ranges from 10^5 to 10^7 per gram of stool in infected individuals.

5.4 OCCURRENCE AND SURVIVAL OF ENTERIC MICROORGANISMS IN LANDFILLS

Few studies have been conducted on the occurrence of enteric microorganisms in landfills or laboratory lysimeters. Kinman et al. (1986) described a 10-year study on the fate of indicator bacteria in 19 large lysimeters maintained by the U.S. Environmental Protection Agency. Large numbers of fecal indicator bacteria were present even after 10 years (Table 4). Because of the large amount of organic matter, some types of enteric bacteria actually may increase in numbers. In another lysimeter study, coliform (but not fecal coliform) bacteria could be detected after 5 years in MSW placed in an indoor lysimeter (Kinman et al., 1986). Donnelly and Scarpino (1983) examined a 9-year-old landfill at various depths (Table 5). The levels of total coliforms, fecal coliforms, and fecal streptococci in all waste materials were relatively high. Results from the solid waste were not consistent with those from leachate samples over the years (data not shown), which showed essentially no indicator organisms present. The higher levels in the top soil may have resulted from animals. Fecal indicator bacteria survived within the landfill, which suggests the possibility that leachate may transport these bacteria to the groundwater if the liner materials were penetrated. A number of pathogenic bacteria, including *Salmonella*, were isolated from the solid waste. The clay liner and the soil beneath the plastic liner, however, showed that few bacterial indicators were present. These last assays demonstrated that the microbe-containing leachate was not able to penetrate the clay liner and soil below to contaminate the groundwater over a 9-year period.

Only one study has looked at the occurrence of enteric viruses in MSW within landfills (Suflita et al., 1992; Huber et al., 1994). In this study, disposable diapers were separated from solid waste from various depths at three different landfills. The diapers had been buried from 2 to 10 years. The 110 diapers containing fecal material were assayed by cell culture for detection of enteroviruses and with gene probes (i.e., nucleic acid hybridization) for poliovirus, hepatitis A virus, and rotavirus. Enteroviruses were not detected in any

TABLE 4. Bacterial Indicators in Solid Waste-Containing Lysimeters

	Wet Refuse (CFU/g)		
Bacteria	**Initial**	**After 2 Years**	**After 10 Years**
Total coliforms	6.8×10^8	5.6×10^4	1.4×10^3
Fecal coliforms	2.6×10^7	5.6×10^3	< 0.2
Fecal streptococci	1.4×10^8	1.6×10^4	2.4×10^2

Source: From Kinman, R. et al. 1986. EPA-600/2-86-035. U.S. Environmental Protection Agency, Cincinnati, OH.

TABLE 5. Fecal Indicator Bacteria Levels in Solid Waste and Leachate From a 9-Year-Old Landfill[a]

Type of Sample	Depth at Which Samples Collected (m)	Total Coliforms	Fecal Coliforms	Total Streptococci
Top soil	0.46	2.4×10^5	3.5×10^3	9.8×10^5
MSW[b]	1.52	1.1×10^2	$<2.0 \times 10^1$	2.0×10^4
MSW	2.13	2.4×10^3	2.0×10^1	6.0×10^4
MSW	2.60	1.6×10^5	2.3×10^2	6.3×10^5
MSW	2.74	5.4×10^4	3.3×10^2	ND[b]
Leachate	3.01	9.2×10^3	3.5×10^2	1.1×10^2
MSW	3.20	3.5×10^4	4.9×10^2	3.5×10^4
Clay liner	3.35	5.0×10^1	$<2.0 \times 10^1$	2.0×10^4
Soil beneath	3.96	5.0×10^1	$<2.0 \times 10^1$	$<2.0 \times 10^4$

[a] Fecal indicator bacteria most probable number (MPN); CFU/100 g or 100 ml.

[b] MSW = municipal solid waste; ND = not done.

Source: Adapted from Donnelly, J.A., and P.V. Scarpino. 1983. EPA-600/2-84-119. U.S. Environmental Protection Agency, Cincinnati, OH.

sample using animal cell culture. Three samples were positive using nucleic acid probes for poliovirus, but negative for rotavirus and hepatitis A. The authors concluded that, although poliovirus RNA was present in some diapers, the viruses were not viable by cell culture assay after 2 years or longer in a landfill. Diapers from the same landfills also were examined for the presence of the protozoan parasites *Giardia* and *Cryptosporidium*. A total of 79 diapers containing fecal material were examined. No *Giardia* cysts were detected in any of the samples, and the only sample positive for *Cryptosporidium* was found in the southeastern U.S. (Rajnarinesingh and Rose, 1991). The diapers were found at a depth of approximately 3.3 m and had been buried for 2 years. Temperatures at the site at which the diaper containing the oocysts was found ranged from 50 to 54°C at a 10-ft depth. At these temperatures, the oocysts would be unlikely to survive for a prolonged period of time (Anderson, 1985). The detection methods used in this study could not determine if the oocysts were still infectious.

Sobsey et al. (1974) studied the survival of poliovirus type 1 and echovirus type 7 in laboratory-scale MSW lysimeters containing simulated refuse. During the 4-month study, temperatures within the lysimeters ranged from 25 to 32°C. Although large numbers of viruses were added to the refuse, none could be recovered from the solid waste after the end of the study period.

In summary, it appears that enteric bacterial indicators may persist and actually grow in landfills. However, the high temperatures generated in landfills cause the rapid die-off or inactivation of pathogenic enteric protozoa and viruses.

5.5 OCCURRENCE AND SURVIVAL OF ENTERIC ORGANISMS IN LEACHATE

A number of studies have looked at the occurrence and survival of enteric microorganisms in landfill leachates since such leachates could contaminate the groundwater in an improperly lined landfill. Long-distance transport of microorganisms, especially viruses, in groundwater has been observed (Keswick et al., 1982). The high concentrations of dissolved solids and organics and low pH which are typical of landfills appear to be antagonistic to the survival of enteric organisms in landfill leachate.

Lysimeter studies indicate that the concentration of indicator bacteria in leachate tends to decline with time, but that low levels may be found in leachate for prolonged periods. Cooper et al. (1975) studied leachate from lysimeters designed to simulate MSW in either open dumps or in sanitary landfills over a 20-week period. The number of fecal streptococci was slightly lower in open dump leachates compared to sanitary landfill leachates; however, the initial numbers were as high as $10^9/100$ ml of leachate. Numbers of coliforms were higher in leachates from open dumps. The coliform concentrations present in the leachates declined relatively rapidly with time in the sanitary landfill configuration, but more slowly in the open dump leachates. Fecal streptococci concentrations did not decline significantly in either configuration, while coliforms and fecal coliforms declined significantly in the leachate from the sanitary landfill lysimeters. Engelbrecht et al. (1974) observed a rapid decrease of total and fecal coliforms in leachate from landfill lysimeters. The initial concentration of fecal coliforms declined from approximately $4.0 \times 10^3/100$ ml to undetectable levels after 100 days of operation. Donnelly (1983) added water to lysimeters once a week to simulate rainfall, and found initial concentrations of total and fecal coliforms in the range of 10,000/100 ml. These concentrations declined to less than 20/100 ml after 20 weeks. Initial fecal streptococci densities of $10^6/100$ ml did not reach undetectable levels until after 1 year. Other studies have observed initial concentrations of total and fecal coliform bacteria as high as 10^5 to 10^6 per 100 ml (Pahren, 1987; Engelbrecht et al., 1974). The observed differences probably reflect differences in addition of stimulated rainfall, make-up of the refuse, packing, time to field capacity, and variation in the toxicity of the leachate to the pathogenic or indicator bacteria.

Several studies also have looked at the occurrence of enteric viruses in leachates. In a study of an experimental landfill cell, Peterson (1972) found poliovirus 3 in 2 out of 13 samples tested over a 31-day period. Sobsey (1978) examined leachates from 21 different landfills in the U.S. and Canada for enteric viruses. All samples were negative except one from an improperly operated landfill where poliovirus types 1 and 3 were detected. Leachates from most of the sites did not contain detectable levels of fecal coliforms, and in only two samples did the concentration exceed $10^3/100$ ml.

Laboratory lysimeter studies by Engelbrecht et al. (1974) and Sobsey et al. (1974) did not demonstrate the presence of enteric viruses in leachates generated by artificially stimulating rainfall events. In these studies, high concentrations of laboratory-grown poliovirus type 1 were added to the refuse in water added to the lysimeter after they were packed with refuse. In both cases the lysimeters were operated for several months. Sobsey et al. (1974) observed that the viruses readily adsorbed to MSW components in the presence of a high concentration of dissolved salts, which are likely to be present in solid waste leachate. In contrast, Cooper et al. (1975) were able to detect poliovirus sporadically in leachates from experimental lysimeters in which the virus had been added to disposable diapers as a fecal slurry at a concentration of $10^6/g$, a level likely to be found in the feces of infected infants. The diapers were placed in lysimeters at different levels as they were filled. Viruses could be detected for up to 166 days in the leachate from one lysimeter. In the study by Cooper et al. (1975), the lysimeters were brought up to field capacity (i.e., water saturation) more rapidly than other studies; this could explain the occurrence of viruses in the leachate. When the lysimeters were brought to field capacity more slowly, no virus was detected. The authors concluded that the rate at which field saturation takes place could be extremely important in predicting the occurrence of viruses in leachates.

The survival of poliovirus, fecal streptococci, and *Salmonella typhimurium* was studied by Engelbrecht et al. (1974) in leachates at pH 5.3 and 7.0 incubated at 22°C; such high temperatures are not uncommon during initial solid waste decomposition in a sanitary landfill. Ambient temperatures of 10 to 20°C were recorded at the periphery of a landfill. The rate of inactivation was greater at pH 5.3 than at pH 7.0; "young" leachate from a landfill is typically between pH 5.0 and 5.5 because of the accumulation of organic acids (see Chapter 2). It was observed that under most conditions *Salmonella* exhibited greater stability in the leachates than the indicator bacteria. Poliovirus type 1 was observed to survive longer than reovirus type 1 in the same leachates.

Glotzbecker and Novello (1975) reported that *Escherichia coli* and *Streptococcus faecalis* survival could vary from hours to months, depending on leachate source and temperature of incubation. They observed that addition of

EDTA to the leachate protected the bacteria against inactivation. Sobsey et al. (1974) also reported that addition of EDTA to landfill leachates was necessary to prevent interference with viral assays. From this evidence it would appear that heavy metals in leachates may interfere with detection of microorganisms and may be responsible for some of the inactivation of these organisms.

Sobsey et al. (1974) found that poliovirus type 1 survival was dependent on the individual leachate sample. A 70% inactivation at 4°C was observed in one sample after 27 days, while a 97% reduction was observed in a second sample. In both leachates more than 95% of the virus was inactivated over 14 days at 20°C, whereas more than 99% was inactivated within 6 days at 37°C.

5.6 FATE OF PATHOGENS DURING COMPOSTING OF MUNICIPAL SOLID WASTE

Composting is a method of stabilizing MSW with the potential of putting it to beneficial use for agriculture or gardening (see Chapter 4). For composting of MSW, removal of glass, metals, and other noncompostable materials is often practiced. General types of composting include controlled reactors, windrows, and aerated piles. There is a great variety of reactor-type composting systems where the fermentation process is carried out in a digester which may be a batch vessel or a continuous system using rotary drums (Haug, 1993). In the windrow system, the refuse is placed in a row (windrow) for periods of 3 to 7 weeks. Air is provided by turning the refuse on a regular basis. In the pile system, air is supplied by drawing air through the compost pile for several weeks. Composting is a thermophilic process, and the temperatures achieved during this treatment (55 to 70°C) are believed to be adequate to kill enteric pathogens. However, for the process to be successful in pathogen elimination, it is important that all of the material undergoing composting be exposed to these temperatures for a sufficient period of time. Cold pockets or zones in nonreactor systems or inadequate resident times in reactor systems may allow pathogens to escape thermal destruction.

5.6.1 Enteric Pathogens

Most of the research on pathogen fate has been on composted sewage sludge. Sewage sludge is likely to contain higher concentrations of enteric pathogens than MSW. Yanko (1989) studied the occurrence of enteric bacteria, viruses, and *Ascaris* eggs in both windrow and static pile composting of sewage sludge. Only viruses were detected during this year-long study of the compost facility. It was concluded that viruses could survive 25 days of composting if the composting mass did not achieve adequately high temperatures. Fecal coliforms also were found in high concentrations if maximum composting temperatures were less than 50°C. Research suggests that temperatures maintained above 53°C for 3 days are sufficient to eliminate enteric pathogens including *Ascaris* eggs (Haug, 1993). It has been found, however, that enteric bacteria can regrow in composted material once temperatures are reduced to sublethal levels. This phenomenon has been reported for total and fecal coliforms, *Salmonella*, and fecal streptococci in liquid sewage sludges. Regrowth may occur through bacteria surviving at undetectable levels or through recontamination from an outside source. Regrowth of *Salmonella* in composted sewage sludges has been noted in several studies (Russ and Yanko, 1981; Hussong et al., 1985; Haug, 1993). Hussong et al. (1985) concluded that the active, indigenous flora of composts act as a barrier to colonization by *Salmonella*. Only in the absence of competing flora, such as by sterilization, were reinoculated *Salmonella* able to regrow.

Limited information is available on the presence of enteric pathogens in composted MSW. Gerba et al. (1995) examined 20 samples of composted MSW for Salmonellae, enteroviruses, *Giardia*, and *Cryptosporidium*. The refuse was first prepared by passing it through a compartmentalized rotating composting vessel, i.e., Eweson digester (Haug, 1993), on a 3- to 4-day batch cycle. The vessel output then was sent to a curing pad for mechanical aeration. During the first 90 days of composting, temperatures ranged from 57 to 70°C within the pile. Samples were collected from days 93 to 203 during the composting period. In some compost piles, the concentration of fecally contaminated diapers had been increased from a normal concentration at this site of 1.6–3.1% to 6.6–7.1%. No protozoan parasites or enteric viruses were detected in any of the samples. *Salmonella* sp. were isolated from one 175-day-old sample in which the diaper concentration had been increased to >7%. Temperatures may not have been uniformly high in that particular pile to kill all of the *Salmonella* or it may have been reintroduced by animals such as birds. Jager et al. (1994) also looked at composting facilities in which the concentration of disposable diapers had been increased to 10%. He did not observe any differences in the isolation of enteric bacterial pathogens in compost with or without the addition of the extra fecally contaminated diapers.

5.6.2 FUNGAL PATHOGENS

Most of the information on the occurrence of *Aspergillus fumigatus* in compost comes from studies on sewage sludge composting. Millner et al. (1977) was the first to study the occurrence of *A. fumigatus* in various stages of aerated static composting. The composting mixture consisted of raw sewage sludge mixed with wood chips as a bulking agent. Wood chips appeared to be the major source of the fungus. Samples of old wood chips contained as many as 6.1×10^7 CFU/g dry weight. Semiquantitative studies of air at the composting site indicated that *A. fumigatus* constituted 75% of the total viable microflora. This decreased to about 2% at a distance of 320 m to 8 km from the site. In a later study, the same authors (Millner et al., 1980) demonstrated that aerosol concentrations from windblown losses from stationary compost piles are relatively small in comparison with those generated during mechanical movement of the piles. After mixing of the compost, *A. fumigatus* concentrations at 3 and

30 m downwind of the static piles were about 33 to 1800 times less than those measured during pile movement and not significantly above background levels. The authors estimated a maximum emission of 4.6×10^6 *A. fumigatus* particles per second during pile moving operations.

An epidemiological study of workers at composting facilities showed that workers did not experience any ill effects in comparison with a control group (Clark et al., 1984). Minor effects were observed, however, which consisted of skin irritation and inflammation of the nose and eyes that resulted from exposure to dust containing *A. fumigatus*. To reduce potential risks to susceptible individuals it has been recommended that personnel working in areas of high agitations such as mixing and screening buildings be evaluated for the development of sensitivity to *Aspergillus* (Haug, 1993).

5.7 CONCLUSIONS

Major gaps in our knowledge of the occurrence and concentration of enteric pathogens in MSW still exist. The relative source of many pathogens in the various components of solid waste has not yet been established. Most of our knowledge of enteric organisms deals with indicator bacteria and not with actual pathogens. Recent development in molecular methods make it possible for the actual detection of enteric pathogens in MSW which was previously not possible or too costly. Information on occurrence allows for a better estimate of pathogen fate in processes used to treat solid waste. In addition, the recent evolution of microbial risk assessment allows for the better decisions which minimize risk from pathogens during solid waste handling, treatment, and disposal.

The fate of many emerging pathogens and their potential occurrence in MSW has yet to be assessed. Every year, new enteric pathogens are discovered and currently recognized ones are given new importance as our ability to document their impact on society increases. For example, hepatitis A virus and enteric adenoviruses are the most thermally stable enteric viruses currently known (Enriquez et al., 1995); yet, little or nothing is known about the occurrence and fate of these viruses in MSW after disposal. Also, the incidence of *Salmonella* infections in the U.S. has increased fourfold in the last 40 years, suggesting we need to know more about the environmental transmission of this pathogenic bacterium. No information appears to exist on the potential for growth of thermophilic pathogens or opportunistic pathogens, such as *Legionella* sp. or *Aeromonas* sp.

While the risks from pathogens in well operated landfills appear to be minimal, it must be realized that the landfill option is becoming less desirable because of environmental regulation. Increased pressure for treatment and recycling brings a need for understanding the fate of pathogens. Each option should be evaluated to minimize environmental risks from pathogenic microorganisms.

REFERENCES

Anderson, B.C. 1985. Moist heat inactivation of *Cryptosporidium* sp. *Am. J. Public Health* 75:1433–1435.

Bennett, J.V., S.D. Holmberg, M.F. Rogers, and S.L. Solomon. 1987. Infectious and parasitic diseases. *Am. J. Prev. Med.* 3:102–114.

Black, R.E., A.C. Dykes, and S.P. Sinclair. 1977. Giardiasis in day-care centers: evidence of person to person transmission. *Pediatrics* 60:486–489.

Clapper, W.E. 1970. Comments on viruses recovered from dogs. *J. Am. Vet. Med. Assoc.* 156:1678–1680.

Clark, C.S., H.S. Bjornson, J. Schwartz-Fulton, J.W. Holland and P.S. Garside. 1984. Biological health risks associated with the composting of wastewater treatment plant sludge. *J. Water Pollut. Control Fed.* 56:1269–1276.

Cooper, R.C., J.L. Potter, and C. Leong. 1975. Virus survival in solid waste leachates. *Water Res.* 9:733–739.

Cruz, J.R., P. Caceres, F. Cano, J. Flores, A. Barlett, and B. Torun. 1990. Adenovirus types 40 and Ead 41 and rotavirus associated with diarrhea in children from Guatamala. *J. Clin. Microbiol.* 28:1780–1784.

Deregnier, D.P., L. Cole, D.G. Schupp, and S.L. Erlandsen. 1989. Viability of *Giardia* cysts suspended in lake, river, and tap water. *Appl. Environ. Microbiol.* 55:1223–1229.

Donnelly, J.A. 1983. Isolation, characterization and identification of microorganisms from laboratory and full-scale landfills. Ph.D. thesis, University of Cincinnati, OH.

Donnelly, J.A., and P.V. Scarpino. 1983. Isolation, Characterization, and Identification of Microorganisms from Laboratory and Full-Scale Landfills. EPA-600/2-84-119. U.S. Environmental Protection Agency, Cincinnati, OH.

Douglas, H., D.S. Reiner, M.J. Gault, and F.D. Gillin. 1988. Location of *Giardia* trophozoites in the small intestine of naturally infected dogs, in P.M. Wallis and B.R. Hammond (Eds.), *Advances in* Giardia *Research*, University of Calgary Press, Alberta, pp. 65–69.

Dudley, D.J., M.J. Ibarra, B.E. Moore, and B.P. Sagik. 1980. Enumeration of potentially pathogenic bacteria from sewage sludges. *Appl. Environ. Microbiol.* 39:118–126.

Emmons, C.W., C.H. Binford, and V.P. Utz. 1979. *Medical Mycology*. Lea & Febiger, Philadelphia, PA.

Engelbrecht, R.S., M.J. Weber, P. Amirhor, D.H. Foster, and D. Larossa. 1974. Biological properties of sanitary landfill leachate, in J.F. Malina, Jr., and B.P. Sagik (Eds.), *Virus Survival in Water and Wastewater Systems,* Center for Research in Water Resources, University of Texas, Austin, pp. 210–217.

Enriquez, C.E., C.J. Hurst and C.P. Gerba. 1995. Survival of enteric adenoviruses 40 and 41 in tap, sea, and waste water. *Water Res.* 11:2548–2553.

Fradkin, L., S.M. Goyal, R.J.F. Bruins, C.P. Gerba, P. Scarpino, and J.F. Stara. 1989. Municipal wastewater sludge. *J. Environ. Health* 51:148–152.

Gerba, C.P., M.S. Huber, J. Naranjo, J.B. Rose, and S. Bradford. 1995. Occurrence of enteric pathogens in composted domestic solid waste containing disposable diapers. *Waste Manage. Res.* 13:315–324.

Gerba, C.P. and J.B. Rose. 1993. Estimating viral disease risk from drinking water, in C.R. Cothern (Ed.), *Comparative Health Risk Assessment*, Lewis Publishers, Chelsea, MI, pp. 117–135.

Gerba, C.P., S.N. Singh, and J.B. Rose. 1985. Waterborne gastroenteritis and viral hepatitis. *CRC Crit. Rev. Environ. Control* 15:213–236.

Glotzbecker, R.A. and A.L. Novello. 1975. Poliovirus and bacterial indicators of fecal pollution in landfill leachates. News of Environmental Research, U.S. Environmental Protection Agency, Cincinnati, OH.

Grew, N., R.S. Gohd, J. Arguedas, and J.I. Kato. 1970. Enteroviruses in rural families and their domestic animals. *Am. J. Epidemiol.* 91:518–526.

Haug, R.T. 1993. *The Practical Handbook of Compost Engineering,* Lewis Publishers, Chelsea, MI.

Howell, R.T. and B.S. Waldron. 1978. Intestinal parasites in Arkansas. *J. Ark. Med. Soc.* 75:212–214.

Huber, M.S., C.P. Gerba, M. Abbaszadegan, J.A. Robinson, and S.M. Bradford. 1994. Study of persistence of enteric viruses in landfilled disposable diapers. *Environ. Sci. Technol.* 28:1767–1772.

Hussong, D.W., D. Burge, and N.K. Enkiri. 1985. Occurrence, growth, and suppression of Salmonellae in composted sewage sludge. *Appl. Environ. Microbiol.* 50:887–893.

Jager, E., H. Ruden, and B. Zeschmar-Lahl. 1994. Communication: microbiological quality of compost with special regard to the input of used panty diapers. *Zbl. Hyg.* 196:245–257.

Keswick, B.H., D.S. Wang, and C.P. Gerba. 1982. The use of microorganisms as groundwater tracers: a review. *Groundwater* 20:142–149.

Kinman, R., J. Rickabaugh, J. Donnelly, J. Nutini, and M. Lambert. 1986. Evaluations and disposal of waste materials within 19 test lysimeters at Center Hoill. EPA-600/2-86-035. U.S. Environmental Protection Agency, Cincinnati, OH.

Lewis, P.D. 1988. Prevalence of *Giardia* sp. in dogs from Alberta, in P.M. Wallis and B.R. Hammond (Eds.), *Advances in* Giardia *Research*, University of Calgary Press, Alberta, pp. 61–64.

Lundgren, D.L., A. Sanchez, M.G. Magnuson, and W.E. Clapper. 1968. Isolation of human enteroviruses from beagle dogs. *Proc. Soc. Exp. Biol. Med.* 128:463–466.

Matsuto, T., and R.K. Ham. 1990. Residential solid waste generation and recycling in the U.S.A. and Japan. *Waste Manage. Res.* 8:229–242.

Millner, P.D., D.A. Bassett, and P.B. Marsh. 1980. Dispersal of *Aspergillus fumigatus* from sewage sludge compost piles subjected to mechanical agitation. *Appl. Environ. Microbiol.* 39:1000–1009.

Millner, P.D., P.B. Marsh, R.B. Snowden, and J.F. Parr. 1977. Occurrence of *Aspergillus fumigatus* during composting sewage sludge. *Appl. Environ. Microbiol.* 34:765–772.

Pahren, H.R. 1987. Microorganisms in municipal solid waste and public health implications. *CRC Crit. Rev. Environ. Control* 17:187–228.

Peterson, M.L. 1972. The occurrence and survival of viruses in municipal solid waste. *Diss. Abstr.* 33/3:2232B–2233B.

Peterson, M.L. 1974. Soiled disposable diapers: a potential source of viruses. *Am. J. Public Health* 64:912–914.

Picking, L.K., W.E. Woodward, H.L. Dupont, and P. Sullivan. 1984. Occurrence of *Giardia lamblia* in day-care centers. *J. Pediatr.* 104:522–526.

Rajnarinesingh, A., and J.B. Rose. 1991. *Cryptosporidium* and *Giardia* from disposable diapers buried in landfills. *Fl. J. Environ. Health* 135:33–39.

Rathje, W. and C. Murphy. 1992. *Rubbish! The Archaeology of Garbage*, Harper Collins, New York.

Rose, J.B. 1990. Occurrence and control of *Cryptosporidium* in drinking water, in G.A. McFeters (Ed.), *Drinking Water Microbiology*, Springer-Verlag, New York, pp. 291–324.

Russ, C.F. and W.A. Yanko. 1981. Factors affecting Salmonellae repopulation in composted sludges. *Appl. Environ. Microbiol.* 41:597–602.

Safferman, R.S., M-E. Rohr, and T. Goyke. 1988. Assessment of recovery efficiency of beef extract reagents for concentrating viruses from municipal wastewater sludge solids by the organic flocculation procedure. *Appl. Environ. Microbiol.* 54:309–316.

Sagik, B.P., S.M. Duboise, and C.A. Sorber. 1980. Health risks associated with microbial agents in municipal sludge, in G. Bitton, B.L. Damron, G.T. Edds, and J.M. Davidson (Eds.), *Sludge — Health Risks of Land Application*, Ann Arbor Science Publishers, Ann Arbor, MI, pp. 15–45.

Sealy, D.P., and S.H. Shuman. 1983. Endemic giardiasis and day-care. *Pediatrics* 72:154–158.

Soares, A.C., T.M. Straub, I.L. Pepper, and C.P. Gerba. 1994. Effect of anaerobic digestion on the enteroviruses and *Giardia* cysts in sewage sludge. *J. Environ. Sci. Health,* A29:1987–1997.

Sobsey, M.D. 1978. Field survey of enteric viruses in solid waste landfill leachates. *Am. J. Public Health* 68:858–864.

Sobsey, M.D., C. Wallis, and J.L. Melnick. 1974. Development of methods for detecting viruses in solid waste landfill leachates. *Appl. Microbiol.* 28:232–238.

Sorber, C.A. and B.E. Moore. 1986. A Critical Review: Survival and Transport of Pathogens in Sludge Amended Soil. EPA/600-S2-87/028. Water Engineering Research Laboratory, Office of Research and Development, U.S. Environmental Protection Agency, Cincinnati, OH.

Suflita, J.M., C.P. Gerba, R.K. Ham, A.C. Palmisano, W.L. Rathje, and J.A. Robinson. 1992. The world's largest landfill: a multidisciplinary investigation. *Environ. Sci. Technol.* 26:1486–1495.

Swabby, K.O., C.P. Hilber, and J.G. Wegrzyn. 1988. Infection of Mongolian gerbils (*Meriones uhguiculatus*) with *Giardia* from human and animal sources, in P.M. Willis and B.R. Hammond (Eds.), *Advances in Giardia Research*, University of Calgary Press, Canada, pp. 75–77.

Ticehurst, J.R. 1991. Identification of hepatitis E virus, in F.B. Hollinger, S.M. Lemon, and H.S. Margolis (Eds.), *Viral Hepatitis and Liver Disease*, Williams and Wilkins, Baltimore, pp. 15–45.

Ungar, B.L. 1990. Cryptosporidiosis in *Homosapiens*, in J.P. Dubey, C.A Speer, and R. Fayer (Eds.), *Cryptosporidiosis in Man and Animals,* CRC Press, Boca Raton, FL, pp. 59–89.

USEPA. 1986. Development of Risk Assessment Methodology for Land Application and Distribution and Marketing of Municipal Sludge. ECAO-CIN-489. Office of Water Regulations and Standards, U.S. Environmental Protection Agency, Cincinnati, OH.

Yanko, W. 1989. Los Angeles Sanitation District, Whitter, CA (personal communication).

6

Testing the Biodegradability of Synthetic Polymeric Materials in Solid Waste

Charles A. Pettigrew and Bradley N. Johnson

CONTENTS

6.1 Introduction 176

6.2 Tiered Testing Approach 179
    6.2.1 Screening Level 180
    6.2.2 Confirmation Level I (Laboratory- and Pilot-Scale Tests)......................180
    6.2.3 Confirmation Level II (Field- or Full-Scale Tests) 180

6.3 Pure Culture Screening Level Tests..180
    6.3.1 Microbial Colonization 180
    6.3.2. Enzyme Assays 182
    6.3.3 Respirometric Analysis Using Pure Cultures 184

6.4 Mixed Culture Screening Level Tests 185
    6.4.1 Clear Zone Assays 185
    6.4.2 Gross Macromolecular Analysis 185
    6.4.3 Respirometric Analysis Using Mixed Cultures 187
        6.4.3.1 Aerobic Test Methods 187
        6.4.3.2 Anaerobic Test Methods 192

6.5 Confirmation Level I (Laboratory- or Pilot-Scale Tests) 193
    6.5.1 Isotopically Labeled Materials...193
    6.5.2 Mesocosm Methods 195
    6.5.3 Residue and Degradation Product Analysis 196

6.6 Confirmation Level II (Field- or Full-Scale Tests) 201
    6.6.1 Biodegradation in Solid Waste Treatment and Disposal...........................201
    6.6.2 Biodegradation in Terrestrial Ecosystems 202

0-8493-8361-7/96/$0.00+$.50
© 1996 by CRC Press, Inc.

6.7 Standardization of Test Methods 203
    6.7.1 Test Conditions...203
    6.7.2 Comparisons Among Tests 203
    6.7.3 Biodegradation Benchmarks 205

6.8 Conclusions 205
    6.8.1 Putting Biodegradation Data to Use..205
    6.8.2 Current Status and Research Opportunities 206

Acknowledgments 207

References 207

6.1 INTRODUCTION

The potential for biodegradation of synthetic polymeric materials, or the lack thereof, has been studied since the 1960s (Darby and Kaplan, 1968; Potts et al., 1973). Since many materials such as plastics, rubber, and textiles are designed to resist biological, physical, and chemical degradation in the environment, synthetic polymers historically have been classified as nondegradable and recalcitrant (Alexander, 1973). The majority of the early research on synthetic polymer biodegradation dealt with issues of biodeterioration as an inherent negative property, rather than the development of materials with relatively short lifetimes. However, as social and political concerns grow about the management of municipal solid waste (MSW), the hazards of plastic litter to wildlife, and plastic film mulching in agriculture, interest in the development of biodegradable synthetic polymers has emerged.

One of the primary features of biodegradable materials is that biodegradation results in a significant reduction in mass of a material. A glossary of terms relating to biodegradation can be found in Table 1. In addition, biodegradable materials are converted to inorganic products that are reassimilated into natural elemental cycles. The combination of mass reduction and recycling of carbon represents an improved environmental profile for biodegradable polymers.

As societies around the world address the potential issues associated with increasing amounts of MSW, costs of disposal, difficulties siting new landfills, and questions about the safety of mass-burn combustion systems, alternative disposal options are being considered. Material recycling, waste-to-energy, composting, and anaerobic digestion of MSW are all expected to increase (Huang, 1990). For example, composting and recycling theoretically are able to treat 60 to 70% of the waste stream (Beyea et al., 1992). During the composting process, much of the organic fraction of the waste stream is biodegraded under controlled conditions and is converted into a product that can be incorporated into the terrestrial environment. The diagram shown in Figure 1 demonstrates the temporal and spatial scales associated with the

TABLE 1   Glossary of Terms

Biodegradation	Defined for the purposes of this chapter as the molecular degradation of a material resulting from the complex actions of microorganisms. The term biodegradation is reserved for the breakdown of organic materials by enzymatic processes leading to the disappearance of the parent molecular structure and to the formation of lower molecular weight organic species, some of which are directly usable for anabolic cell reactions; catabolized to CO_2, CH_4, and water; or incorporated into stable humic materials.
Primary biodegradation	Occurs when an initial, small alteration is made to the molecule, changing its physico-chemical properties and physical integrity. This usually happens during the first steps of the biodegradation pathway. Primary biodegradation can be seen merely as the first step of biotransformation of a molecule that often results in the formation of metabolites or residues.
Ultimate biodegradation	Occurs when the chemical material is broken down and all the organic carbon is converted to CO_2 or CH_4 and/or is incorporated into biomass or stable humic materials. Ultimate biodegradation thus leads to the complete conversion of organic carbon with extensive mineralization and no persistent metabolites.
Inherent biodegradability	Means that a material can be ultimately biodegraded under a certain set of circumstances. In principle, inherent biodegradability is related to the intrinisic properties of the chemical. It also can be intended as the maximum level of biodegradation that a chemical may reach under optimum conditions and extensive exposure times.
Practical biodegradability	Means that a material is ultimately biodegraded in the environment in which it is disposed. Furthermore, the biodegradation rate of a practically biodegradable material exceeds or is similar to the loading rate of the chemical into the environment such that biodegradation is a meaningful removal mechanism, and the chemical does not accumulate.

various stages of polymer biodegradation in a compost/soil environment. It is important to recognize that polymer biodegradation can be studied at either the macro or molecular level and that information about both is usually required to understand or predict the biodegradability of a material.

Accurate assessment of the biodegradability of polymeric materials in solid waste treatment and disposal systems and in terrestrial environments is often difficult, because numerous inherent chemical factors as well as environmental factors can affect the biodegradation profile of polymeric compounds. Table 2 provides a list of many of the key inherent chemical properties of a polymer, as well as environmental factors, that can affect polymer biodegradation. As

Macro Scale

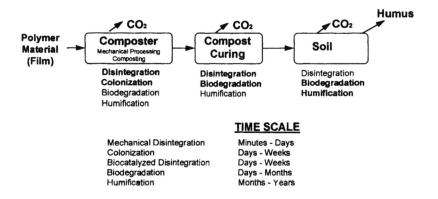

Molecular Scale

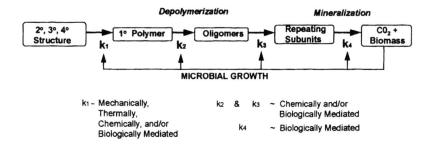

FIGURE 1   Diagrammatic representation of the macro- and molecular scales associated with polymer biodegradation in a compost/soil environment (From Federle, T.W., unpublished material.)

a consequence of these chemical and environmental factors, the determination of the biodegradability of a polymer can be highly dependent on the test system and methods used to make the assessments. While numerous methods have been developed to test the biodegradability of natural and synthetic polymers, the lack of consistent standard methods often has resulted in confusion and misinterpretation among scientists and the lay public (Palmisano and Pettigrew, 1992; Yabannavar and Bartha, 1994). It is important to understand the specific advantages and limitations of test methods to standardize testing procedures and to generate data with good predictive value.

The purpose of this chapter is to review the literature on test methods used to assess the biodegradability of synthetic polymeric materials in the context

TABLE 2 Examples of Factors Affecting Polymer Biodegradation

Inherent Properties	Polymer Example	Ref.
Molecular weight	Polyethylene glycol	Kawai et al., (1978)
Bond type	Lignocellulose	Crawford (1981)
Substitution type	Cellulose acetate/propionate	Komarek et al., (1993)
Degree of branching	Aliphatic polyesters	Potts et al., (1973)
Degree of crosslinking	Polycaprolactone, rubber	Jarret et al., 1982, Tsuchii et al., (1990)
Degree of crystallinity	Polycaprolactone	Benedict et al., (1983a,b)

Environmental Properties		
Presence of degraders	Poly (beta-hydroxybutyrate)	Nishida and Tokiwa (1993)
Temperature	Polycaprolactone	Pettigrew et al., (1995)
Nutrients	Lignin	Kirk et al., (1976)
Moisture	Cellulose acetate	Gu et al., (1994)
Inhibitors	Rubber	Dimond and Horsfall (1943)
Sorption	DNA	Lorenz and Wackernagel (1987)
Adaptation	Polycaprolactone	Kavelman and Kendrick (1978)

of MSW treatment and disposal. In addition, we describe a tiered testing approach for evaluating polymer biodegradability and provide examples of how commonly used test methods fit into this tiered approach. The intent is to elucidate key concepts and techniques that should be considered in determining and predicting synthetic polymer biodegradability during MSW treatment and disposal and in terrestrial environments.

6.2 TIERED TESTING APPROACH

Understanding synthetic polymer biodegradation requires knowledge of the diverse biotic and abiotic factors that interact to facilitate or limit polymer biodegradation. MSW treatment systems and terrestrial environments are very heterogeneous. Spatial gradients and temporal variability are observed for many properties, including temperature, moisture, pH, organic matter, inorganic nutrients, biomass quantity and diversity, biological activity, and oxygen availability. Because of this complexity, model laboratory systems typically have been used to control variables, normalize the spatial heterogeneity, and shorten the temporal scales. Many of these controlled laboratory systems generate important information about the mechanisms of synthetic polymer biodegradation, but they often result in data with limited value in predicting the environmental fate of a material. Ultimately, all laboratory test methods must be validated and verified through more complex, environmentally realistic testing and through experience gained from measurement of the practical biodegradability of synthetic polymers.

This chapter presents polymer biodegradation test methods in three basic levels that comprise a "tiered testing" approach. This tiered testing approach has proven useful in the evaluation of new synthetic polymers and is used as the outline for organizing this review of existing polymer biodegradation methods.

6.2.1 Screening Level

At this level, questions about the inherent biodegradability of a synthetic polymer are addressed. Experiments are done under controlled laboratory conditions to ensure reproducibility and precision, but, because the conditions often are optimized, observed biodegradability may not correlate with that of "real world" disposal. The intent is to gather preliminary data.

6.2.2 Confirmation Level I (Laboratory- and Pilot-Scale Tests)

At this level, questions about the practical biodegradability of a synthetic polymer are often addressed. Thcsc questions are addressed by conducting experiments under controlled laboratory- or pilot-scale conditions, but special attention is paid to design experiments to generate information that predicts whether or not a synthetic polymer will biodegrade under "real world" conditions.

6.2.3 Confirmation Level II (Field- or Full-Scale Tests)

The expected *in situ* biodegradation of a synthetic polymer during solid waste treatment (compost facilities) and disposal (landfill facilities) and in terrestrial environments is addressed at this level. Demonstration of *in situ* biodegradation is intended to ensure that the parent polymeric material or persistent intermediates or residues are not accumulating in the environment over time to levels that could cause adverse environmental consequences.

6.3 PURE CULTURE SCREENING LEVEL TESTS

Methods used to assess the biodegradability of organic materials are developed with a biochemical, organismal, or ecological perspective. Biochemical and organismal methods that utilize axenic or pure cultures of microorganisms have been used widely to assess the inherent biodegradability of synthetic polymers and to gather information regarding mechanisms of polymer biodegradation. Examples of these methods include microbial colonization, enzymatic depolymerization, and manometry.

6.3.1 Microbial Colonization and Growth

When challenged with characterizing the biodegradation potential of a polymer, many investigators have used screening level methods that focus on the isolation and use of specific microorganisms that are capable of metabolizing the polymer in some way. The initial choice of microorganisms in these studies often has been based on experience with polymer biodeterioration research (Osmon and Klausmeier, 1978). The source of microbial strains may be environments with higher than normal populations of organisms capable of degrading the polymer as the result of previous exposure. For example, Tsuchii et al. (1985) isolated a *Nocardia* species from samples of deteriorated rubber buried in soil. In other cases, isolation of strains may be achieved from an environmental matrix containing a diverse array of microorganisms. For example, soils and composts were found to be good environments to enrich for microorganisms that biodegrade polymeric materials such as polycaprolactone (PCL) (Kavelman and Kendrick, 1978) and polyhydroxyalkanoates (Gilmore et al., 1992).

Rapid and reproducible methods have been developed to screen natural materials (e.g., rubber, leather, hydrocarbons) for their susceptibility to fungal and bacterial decay. These methods are equally useful as screening tools in assessments of the inherent biodegradability of polymers. Test materials are surrounded, but not covered, by molten agar, inoculated with a select fungal or bacterial strain, and incubated for up to 21 days, usually at mesophilic temperatures (ASTM, 1991). Microbial growth is then rated qualitatively as the percentage of the polymer specimen covered, the abundance of mycelia, or colony growth. Resistance of the material to decay is qualitatively inferred by comparing test samples to control materials known to be either highly resistant or susceptible. Some versions of this method (ASTM, 1990, 1991) prescribe the use of specific fungal species, such as *Penicillium pinophilum, Chaetomium globosum, Gliocladium virens,* and *Aureobasidium pullulans.* Other test methods (ASTM, 1980) utilize bacterial isolates such as *Pseudomonas aeruginosa.* The white rot fungi *Phanerochaete chrysosporium, Trametes versicolor,* and *Pleurotus ostreatus* are also commonly used (Milstein et al., 1994). Using similar pure culture assays, Benedict and coworkers (1983a,b) recommended that several species of fungi and bacteria be used in screening tests, thus ensuring broad representation of microbial biodegradation capabilities.

The majority of studies that have used microbial colonization assays have relied primarily on the growth rating as an endpoint. In many cases, the polymeric material can be recovered at the end of the test and used for additional analyses such as mass loss and changes in physical or chemical properties. Liquid media are sometimes used in place of agar media to facilitate recovery and to bypass problems associated with trace organic contaminants in agar. Lee et al. (1991) found that changes in tensile strength, percent elongation, and molecular weight distribution of polymeric samples could be measured in microbial colonization assays. Weight loss data were inconclusive, however, due to cell mass accumulation on the sample. While biological attack can result in a decrease in molecular weight or change in tensile strength, it does not always lead to solubilization or ultimate biodegradation. Similarly, changes in tensile strength and percent elongation can result from biological attack.

In summary, microbial colonization assays offer relatively rapid and inexpensive methods for assessing the inherent biodegradability of synthetic polymers. Since these methods are conducted under controlled conditions and key microbial strains are available from public culture collections, these tests generally are reproducible from one laboratory to another. Use of the growth-rating endpoint tends to result in qualitative data, but more quantitative data can be obtained with alternative physical or chemical property endpoints. False-negative results can occur when growth of a microbial population requires additional nutrients (e.g., vitamins) that the test substance does not provide, or when the microorganisms capable of biodegrading the polymer are not present in the prescribed test. Likewise, false-positive results occur when carbon sources other than the test material (e.g., plasticisers) serve as the growth substrate.

6.3.2 Enzyme Assays

The identification of microorganisms that degrade a synthetic polymeric material and the subsequent characterization of the enzymatic mechanisms that are employed by the microorganisms are fundamental to a mechanistic understanding of polymer biodegradation. While detailed studies of some naturally occurring polymers and their derivatives, such as cellulose and starch (Seal, 1988) and polyhydroxyalkanoates (Doi, 1990) are available, the processes of enzymatic depolymerization of synthetic polymers are generally poorly understood. In cases where microorganisms are known to produce depolymerase enzymes that show activity against a specific type of polymer chemistry, purified enzyme preparations can be used to study the effects of variations in the polymer substrate. These are useful in screening-level biodegradation tests. For example, various stereoisomers of poly (beta-hydroxybutyrate) (PHB) have been screened for degradability using purified polyhyydroxyalkanoate depolymerase enzymes from several bacterial and fungal isolates (Hocking et al., 1994; Abe et al., 1994).

While knowledge of the monomer subunits of a polymer provides some insight into the macromolecular structure and susceptibility to enzymatic attack, there are often several different ways that the monomers can be arranged. In the case of PHB, the methyl side group of bacterial PHB is always positioned in the same direction with respect to the polymer chain and is described as having an isotactic configuration (Abe et al., 1994). Recently, synthetic PHB stereoisomers have been prepared with different stereoregularities (Hocking et al., 1994). In the syndiotactic form, the methyl groups alternate positions down the PHB chain. The results shown in Figure 2 were generated with film samples of PHB containing varying amounts of isotactic diad fractions [i]. The enzymatic erosion, measured as weight loss, of PHB films ranging in [i] value from 0.68 to 0.92 were higher than that of poly [(R) 3-hydroxybutyrate] films prepared from bacterial PHB ([i] = 1.00) as well as synthetic PHB samples that were syndiotactic. These data suggest that some

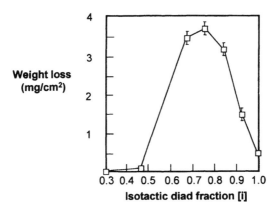

FIGURE 2 Effect of stereoregularity on the weight loss of poly (beta-hydroxybutyrate) films after enzymatic degradation for 3 h with poly (beta-hydroxybutyrate) depolymerase (2.0 µg/ml) from *Alcaligenes faecalis* at 37°C. The weight loss data were averaged on three film samples. (From Abe, H., I. Matsubara, Y. Doi, Y. Hori, and A. Yamaguchi. 1994. *Macromolecules* 27:6024. Copyright 1994, American Chemical Society. With permission.)

synthetic forms of PHB will biodegrade faster than bacterial PHB. The assay used to generate the data presented in Figure 2 provides a rapid measure of the inherent biodegradability of PHB stereoisomers.

Lignin provides a good case study of the role of enzyme activities in polymer decomposition. Lignin is highly cross linked, being formed by enzymatic and free radical copolymerization of coumaryl, guaiacyl, and syringyl alcohols (Kirk and Farrell, 1987). Although lignin is resistant to rapid decomposition, lignin is known to be degraded by a number of enzymes produced by white rot fungi and actinomycetes. Among these are the oxidative enzymes laccase and quinone oxidoreductase produced by strains of *Polyporus versicolor* and *Chrysoporium lignorum* (Westermark and Eriksson, 1974); manganese(II) peroxidase produced by *Phanerochaete chrysosporium, Pleurotus ostreatus*, and *Trametes versicolor* (Milstein et al., 1992); and lignin peroxidase produced by *Phlebia radiata, Panus tigrinus,* and *Bjerkandera adusta* (Kirk and Farrell, 1987). Laccase catalyzes the one-electron oxidation of phenols to phenoxy radicals (Reinhammer, 1984). Manganese peroxidase and lignin peroxidase are thought to function as phenol oxidizing enzymes by introducing functional groups into the lignin structure (Kirk and Farrell, 1987; Milstein et al., 1992). Enzyme assays use small quantities of test material and reagents and provide a rapid assessment of many different polymer types.

Degradation pathways of complex polymers such as lignin can be elucidated using enzymatic techniques. For example, Tien and Kirk (1983) isolated an oxidative, H_2O_2-requiring enzyme from *Phanerochaete chrysosporium* cultures and applied it to reconstituted model lignin substructures. They used thin layer chromatography to resolve reaction products following incubation and were able to deduce specific sites of enzymatic attack. Cleavage of specific C–C bonds was demonstrated for the two lignin substructures 1,2-diarylpropane and arylglycerol-beta-aryl ether. In a similar manner, Wojtas-Wasilewska et al. (1988) isolated protocatechuate 3,4-dioxygenase from *Pleurotus ostreatus* and immobilized the enzyme on controlled-porosity glass beads. Fractions of sodium-lignosulfonate were then exposed to the enzyme and characterized by ultraviolet spectroscopy. These investigators were able to document the degradation of aromatic constituents of the lignin substrate.

Plant litter biodegradation is largely mediated and controlled by extracellular enzymes. While good fundamental knowledge of degradation mechanisms for some naturally occurring macromolecules such as lignin and lignocellulose is available, the application of enzymatic methods to ecological studies of biodegradation has been limited (Sinsabaugh et al., 1991). Some studies have assessed the biochemical and kinetic characteristics of enzymes in soil and have attempted to relate these data to key environmental events such as the biodegradation of xenobiotic compounds (Burns, 1982). These ecological studies are very critical in determining the role of enzymes in the overall degradation of polymers in waste treatment and the environment (Adney et al., 1989). Enzymes become better quantified and understood, it is likely that enzyme methods will be used more frequently.

6.3.3 Respirometric Analysis Using Pure Cultures

Many investigators have used specific microorganisms in screening-level methods that rely on respirometric techniques to measure the consumption of oxygen during the aerobic biodegradation of synthetic polymeric materials. Burgess and Darby (1964) provided early evidence that manometric techniques can provide rapid and accurate measurement of the microbial biodegradation of plastic materials. However, Tilstra and Johnsonbaugh (1993) noted that the success of manometric techniques depends on the choice of microbial species and the growth medium, as well as the purity of the polymer material. For example, Burgess and Darby (1964) found that the weight loss of plastic films was well correlated with oxygen consumption but that the observed weight loss was due to the biodegradation of plasticizers present in the material and not due to polymer biodegradation.

The screening level methods described above have been limited to cases where microbial isolates can directly metabolize test polymers as a carbon or energy source. However, there are other situations where polymer degradation occurs by the action of exogenous enzymes in conjunction with enzymes produced by other organisms or when specific nutrients become limiting. The absence of these additional enzymes or the appropriate nutrient conditions in screening level tests may lead to false-negative assessments of the capability of an isolate to degrade a material. Kirk et al. (1976) observed that growth of *Coriolus versicolor* in culture with lignin as the sole carbon source was negligible, but lignin degradation increased when cellulose, xylose, glucose, or cellobiose were added. The degradation of humic substances extracted from soil also has been shown to be affected by medium conditions. In particular, the ability of *Phanerochaete chrysosporium* to degrade natural humic substances is dependent upon the presence of glucose, can be increased by nitrogen limitation in culture (Blondeau, 1989), and requires a pH less than 6.5 (Kontchou and Blondeau, 1992). The degradation of humic substances was the result of secondary metabolism, attributed to extracellular activity of a relatively nonspecific lignin peroxidase enzyme. This type of substrate interaction illustrates why simple screening tests cannot adequately predict the biodegradation of polymers under realistic conditions. Their usefulness is limited to developing presumptive evidence for establishing biodegradation mechanisms.

6.4 MIXED CULTURE SCREENING LEVEL TESTS

6.4.1 Clear Zone Assays

An additional agar-based screening tool, the clear zone method, is a modification of the microbial colonization or growth rating assay methods described above. This methodology was developed to isolate polymer degrading strains from mixtures of microorganisms typically found in environments such as compost or soil. Microorganisms capable of degrading a polymeric material are determined by looking for zones of clearing around microbial colonies on agar media that is opaque due to the presence of powdered polymer. This method is demonstrated by the work of Gilmore et al. (1992) and Nishida and Tokiwa (1993) where microbial isolates with aliphatic polyester biodegradative capabilities were obtained from mixed microbial populations in compost and soil samples. The clear zones surrounding the microbial colonies shown in Figure 3 were obtained by inoculating a compost sample on a nutrient agar medium containing emulsified PHB. While the appearance of a clear zone around a microbial colony does provide initial evidence for polymer biodegradation, this screening method should be followed by more rigorous tests of inherent biodegradability as presented below.

6.4.2 Gross Macromolecular Analysis

In many *in vitro* tests with synthetic polymers, gross macromolecule analyses are used to evaluate the susceptibility of natural materials to microbial attack; however, it should be recognized that few of these methods are able to quantify complete biodegradation of the polymer. Among these analyses, weight loss is most widely used (ASTM, 1986), based on the theory that a decrease in recoverable material from a test system is an indication of disintegration or complete mineralization. Weight loss is typically expressed on a dry weight basis as follows:

$$\text{weight loss } (\%) = [(R_1 - R_2)/R_1] \times 100$$

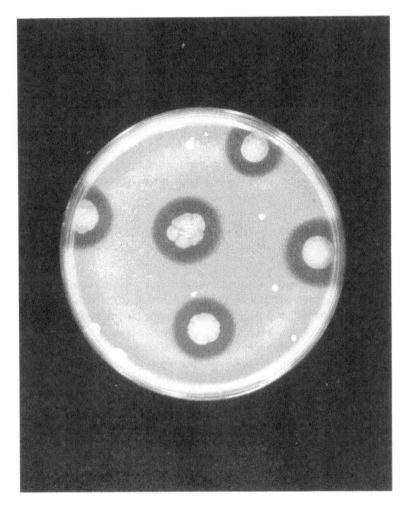

FIGURE 3   Microbial colonies and clear zones on a nutrient agar plate containing emulsified polyhydroxybutyrate. Sample was obtained from yard-waste compost and cultured at 30°C for 2 days. (From Goodwin, S.D., unpublished material).

where R_1 and R_2 are oven dry weights of initial (fresh) and final (exposed) samples. Weight loss measurements for exposed polymer films often are normalized to the exposed surface area, taking into account both sides of the film (Gu et al., 1993). For materials such as vulcanized rubber, volatilization losses of components may be avoided by drying materials *in vacuo* (Tsuchii et al., 1990).

Scanning electron microscopy (SEM) is useful in assessing visual changes in the surface morphology and can be used to infer microbial attack on synthetic polymeric materials. For example, Gilmore et al. (1992) used this technique to observe grooves and pits in exposed plastics films, and Watanabe (1985) used SEM to make visual observations of the stages of attack on acrylic textile fibers by *Penicillium notatum* and *Aspergillis niger*. Erosion of fiber surfaces was seen to proceed via splits along the axis of the fiber, with actual fiber breakage representing a final stage of physical change. Colonization by microorganisms also can be observed (Gu et al., 1993), providing a qualitative evaluation of the general affinity of organisms for the material. Sample preparation is minimal, with material samples recovered from composts or soil washed to remove coarse debris and dried *in vacuo* before being coated with gold (Gilmore et al., 1992). It should be noted that SEM is of little use in determining specific causes for changes in surface morphology, and the incidence of surface-attached microbes cannot be used to infer biodegradation, since microbes may merely be sorbed onto the surfaces and may be metabolizing substrates other than the polymer itself.

For synthetic materials exposed to microbial attack from pure or mixed cultures, other physical and visual observations include: (1) surface overgrowth; (2) discoloration/change in opacity; (3) change in tensile strength, stiffness or hardness; (4) change in water vapor transmission; and (5) intrinsic viscosity of polymer dissolved in solvent. For example, Williams (1982) investigated rubber biodegradation by dissolving rubber samples in toluene and allowing the solutions to flow through capillary tubes to measure intrinsic viscosity. Efflux times of degraded rubber samples were compared to those of standard solutions of polymers of known molecular size. This methodology provided an indication of polymer biodegradability by using intrinsic viscosity measurements to estimate changes in polymer molecular weight. In most cases, comparisons of molecular weight between exposed and fresh samples will provide indirect evidence of degradation (Borel et al., 1982; ASTM, 1990; Gilmore et al., 1992). Changes due to biotic attack may be inferred by comparison with control samples incubated in the absence of microorganisms. This can be achieved by surface sterilizing samples (e.g., with ethanol) and autoclaving solid or liquid growth media.

6.4.3 Respirometric Analysis Using Mixed Cultures

The degree and rate of biodegradation of synthetic polymeric materials can be determined using mixed populations of microorganisms in screening level tests that rely on respirometric techniques to measure O_2 consumption or the production of CO_2 or CH_4 associated with aerobic or anaerobic mineralization of the polymer, respectively. The list of test methods shown in Table 3 demonstrates the types of respirometric techniques currently available for polymer biodegradation testing. The basic design and key features of these methods are discussed below.

6.4.3.1 Aerobic Test Methods

The biodegradability of a polymer can be determined by measuring O_2 consumption by microbial populations due to the oxidation of a polymeric test material. The Closed Bottle test is a variation of the standard biochemical

**TABLE 3   Standard Respirometric Test Methods Used
              To Assess Polymer Biodegradation**

Test Method	Parameter	Ref.
Aerobic		
Closed bottle test	O_2 uptake	OECD (1981)
Automated respirometer	O_2 uptake	OECD (1981), ASTM (1993)
Sturm test	CO_2 production	OECD (1981), ASTM (1992a)
Compost-CO_2 test	CO_2 production	ASTM (1992b)
Soil-CO_2 test	CO_2 production	OECD (1981), USFDA (1987)
Anaerobic		
Biochemical methane	CO_2 and CH_4 production	Shelton and Tiedje (1984),
potential (BMP)		CSA (1992), ASTM (1992c)

oxygen demand (BOD) test that has been widely used in sewage treatment control (APHA, 1989). Oxygen consumption is monitored over time vs. a control sample that contains no test material. The results of the Closed Bottle test are expressed as the theoretical O_2 demand that could be consumed if all of the carbon in the test material is oxidized to CO_2. Current versions of the Closed Bottle test, designated 301D by the Organization for Economic Cooperation and Development (OECD, 1981), use an exposure time of 28 days instead of the traditional 5 days used in the BOD test. The Closed Bottle test uses a mineral salts solution that fills the container, leaving no gas headspace. Therefore, the initial O_2 concentration is limited to 8–10 mg/l and care must be taken to avoid adding too much test material or too much biomass and depleting the O_2 supply. Fischer (1971) noted that use of pure O_2 instead of air to saturate the starting test medium will allow greater O_2 dissolution and the use of higher test substance concentrations. One problem with methods that monitor O_2 removal is that microorganisms mediate nitrification and the oxidation of other noncarbon components in the test matrix. One solution is to add nitrification inhibitors such as 2-chloro-6-(trichloromethyl)pyridine or N-allylthiourea to the medium (APHA, 1989). Krupp and Jewell (1992) noted that extending the test duration to 60 days enhanced the measurement of plastic film biodegradation. In similar experiments, van der Zee et al. (1994) found that 42-day experiments were sufficient for biodegradation testing of starch-based plastics in the Closed Bottle test.

Using the same O_2 consumption principle, more sophisticated BOD tests have been developed and used as standard methods. These automated tests, designated 301C by the Organization for Economic Cooperation and Development (OECD, 1981) and D 5271-93 by the American Society for Testing and Materials (ASTM, 1993), make use of BOD meters or respirometers to measure the O_2 demand continuously during the course of a biodegradation experiment. These methods avoid some of the problems associated with the limited oxygen supply in the Closed Bottle test by allowing O_2 to be replaced in the gas head space. Since O_2 is not limiting, these methods allow a higher initial test substance concentration and higher biomass inoculation. Using the OECD 301C method, Yakabe and Tadokoro (1993) found that a higher biomass inoculum and smaller test substance particle size facilitated the rate of bio-degradation of PCL. However, these investigators also noted that particle size and biomass concentration did not affect the biodegradability of a microbial polyester, poly (3-hydroxybutyrate-co-3-hydroxyvalerate) (PHB/PHV). Yak-abe and Tadokoro (1993) speculated that the difference in depolymerization of PHB/PHV and PCL was due to the difference in activities of the respective depolymerase enzymes.

A more direct measure of the aerobic biodegradation of synthetic polymers is obtained by measuring CO_2 production, since CO_2 is a direct biodegradation end product. The Sturm test is an example of a screening-level test that measures CO_2 production as the biodegradation endpoint (Sturm, 1973). Test compounds are placed in an inorganic medium and inoculated with the super-natant of homogenized activated sludge-mixed liquor (typically 1% v/v). After the addition of test compound, CO_2 production is monitored over time and compared to a control sample that contains no test material. The Sturm test measures the mineralization of the test material, and the results are expressed as the percentage of theoretical CO_2 that could be produced if all of the carbon in the test material was mineralized to CO_2. The Sturm test was developed initially to assess the biodegradability of relatively low molecular weight materials that are disposed via sewers. Due to the paucity of alternative meth-ods, however, the Sturm test often has been used by default for testing the biodegradability of high molecular weight polymers (ASTM, 1992a). The Modified Sturm test, designated 301B by the Organization for Economic Cooperation and Development (OECD, 1981), discontinued the pre-adaptation procedure for the inoculum that was developed by Sturm (1973) to make the test more stringent.

The results presented in Figure 4 demonstrate that the conditions of the Sturm test are not optimized for polymeric materials and that several limita-tions must be considered when interpreting Sturm test data. First, the Sturm test originally was designed to run for 1 month. Polymer biodegradation studies often must be run longer than 1 month, and microorganisms may die off over time so that reinoculation is required. Second, while an activated sludge inoculum is very diverse, it is composed primarily of bacterial populations, and the conditions of the Sturm test are not conducive to growth of filamentous microorganisms (e.g., actinomycetes and fungi). These groups of microorgan-isms are known to play key roles in the biodegradation of natural polymeric materials such as plant litter. One possible solution to this limitation could be achieved by inoculating with a soil or soil extract instead of activated sludge. Third, the Sturm test was designed to assess single chemicals. Many polymeric materials consist of multiple chemicals. For example, the results presented in Figure 4 demonstrate that plant litter materials composed of complex assem-blages of heterogeneous polymers will degrade much differently from a homopolymer such as cellulose.

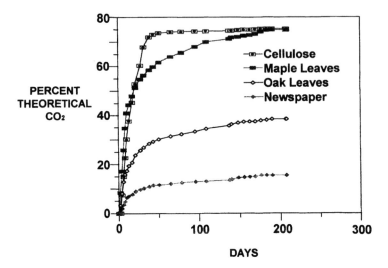

FIGURE 4 Biodegradation of various cellulosic materials in the modified Sturm test. (As described by the Organization for Economic Cooperation and Development. 301-B. 1981.)

Because of the Sturm test limitations, alternative screening-level tests that also measure CO_2 production as the biodegradation endpoint have been developed. Alternative screening level test methods have been designed to simulate MSW disposal conditions as well as terrestrial environments. These methods involve the use of high biomass matrices (i.e., compost or soil) where the optimum CO_2 signal-to-noise ratio is maintained by increasing the test substance concentration. The Compost-CO_2 test is an aerobic biodegradation test that uses stabilized mature compost derived from the organic fraction of MSW as the microbial inoculum and the test matrix. This test simulates an intensive aerobic composting process via a prescribed temperature regime and has been designated D5338-92 by the American Society for Testing and Materials (ASTM, 1992b). The incubation temperatures specified by the D5338-92 method are day 0–1, 35°C; day 1–5, 58°C; day 5–28, 50°C; day 28–45, 35°C. The D5338-92 method facilitates the exposure of the test material to a broad diversity of microorganisms in a high biomass system, but the high test substance concentration (10%, w/w) required to achieve good signal-to-noise ratios can be problematic (Pettigrew et al., 1995). In addition, Pettigrew et al. (1995) noted that the artificially imposed temperature regime causes drastic shifts in microbial activity that may affect results. Gu et al. (1993) have avoided some of these shortcomings in a Compost-CO_2 test that used a simulated compost matrix (to facilitate reproducibility) and lower test substance concentrations (0.5%, w/w) and maintained a constant temperature at 53°C.

Soil biodegradation testing is done most commonly under aerobic conditions using CO_2 production as the biodegradation endpoint. The inherent biodegradability of a material is determined in a closed container, often referred to as a Biometer flask, containing a well characterized soil sample. As for the Sturm test, the CO_2 that is generated is trapped in an alkaline solution and is subsequently titrated or measured in a carbon analyzer. Radiolabeled samples can be measured by scintillation counting (see below). Standard soil biodegradation test methods, exemplified by the U.S. Food and Drug Administration Environmental Assessment Technical Assistance handbook (USFDA, 1987), originally were selected because of their relative simplicity and broad applicability (Sharabi and Bartha, 1993). Yabannavar and Bartha (1994) found these soil biodegradation test methods to be useful screening-level tools for assessing the biodegradability of plastic films in soils. However, these investigators noted that, for many plastic materials, positive results must be verified by other analytical methods as CO_2 production may result from the degradation of additives or single components of a composite material.

An interesting combination of test methods, recently proposed to the American Society for Testing and Materials D20.96 subcommittee on environmentally degradable plastics, involves determining the further conversion of the undegraded or partially degraded test material following a Compost-CO_2 test (ASTM, 1992b) by bringing the compost-borne material in contact with soil. In most cases, polymeric materials are not expected to be completely mineralized during composting, and it is important to assess further mineralization after the compost is applied to soil. The combination of a Compost-CO_2 test followed by a soil biometer test may provide a better understanding of the terrestrial fate of polymers than the results of either test individually.

Compared to the Sturm test, solid matrix methods provide more environmentally relevant conditions, especially for solid waste-borne materials; however, these methods also have limitations. For example, while testing several fundamental assumptions about soil biodegradation test methods, Sharabi and Bartha (1993) found that greater than half of the net CO_2 evolution in soil biodegradation tests may represent the mineralization of biomass and soil organic matter that is not related to the test material. Typically, this "background" mineralization is accounted for by using unamended control data (subtracted from the test system response). In screening level tests using unlabeled materials, the addition of high concentrations of unlabeled test substance is required to overcome the limitations of a high background mineralization. Sharabi and Bartha (1993) noted these high substrate conditions can stimulate the mineralization of soil organic matter and contribute to CO_2 evolution above background controls (i.e., net CO_2). This potential "priming effect" demonstrates that while net CO_2 evolution measurements are useful in first tier, screening-level tests for assessing the inherent biodegradability of substrates in compost or soil environments, these first tier tests may not accurately predict the true mineralization of the substrate in solid waste treatment and terrestrial environments. Use of ^{14}C-labeled test materials that are dosed at realistic concentrations can avoid this limitation.

Sorption onto solids is another factor that may make polymers or polymer biodegradation products unavailable to microbial or enzymatic attack in a soil system, as demonstrated by Lorenz and Wackernagel (1987) for DNA. Adsorbed molecules are not generally thought to be available for intracellular uptake. While the adsorption of polymers such as DNA to well characterized sorbents such as sands and pure clays has been thoroughly studied, studies of polymer and oligomer adsorption to complex and heterogeneous soils have only recently been initiated (Ogram et al., 1994). The adsorption of polymers and its effect on biodegradability require careful consideration of the soil mineral content, solution composition, and organic content of the soil as well as the inherent chemical properties of a specific polymer.

6.4.3.2 Anaerobic Test Methods

Anaerobic conditions are found in MSW landfills, anaerobic digesters for MSW and sewage sludge, and some soil and sediment environments. The anaerobic biodegradation of polymeric materials by bacterial consortia results in the conversion of carbon into CO_2 and/or CH_4. Making use of this knowledge, Owen et al. (1979) developed the screening-level Biochemical Methane Potential (BMP) assay for assessing the anaerobic biodegradability of various organic materials. Shelton and Tiedje (1984) modified this BMP test approach to develop a rapid and inexpensive method for determining the anaerobic biodegradation of a wide variety of materials, including synthetic polymers. This test method uses a mineral salts medium inoculated with anaerobic digester sludge, and the amount of CO_2 and CH_4 produced is determined via pressure transducers that continuously measure the pressure in a flask. Gas samples are analyzed periodically by gas chromatography to determine the composition of gas in the head space of the flask. The BMP test was designed to operate in the mesophilic temperature range (i.e., approximately 35°C)

Wang et al. (1994) used the BMP test method to assess the anaerobic biodegradability of cellulose and hemicellulose in excavated refuse samples. They determined that the theoretical gas production calculated from measured cellulose and hemicellulose concentrations was not well correlated with measured gas values. These investigators noted that the lignin content of the refuse probably played a role in limiting the amount of cellulose available for decomposition in this assay. Likewise, Day et al. (1994) used a similar Canadian Standard Association test method, designated CAN/CSA-Z218-M (Canadian Standard Association, 1992), using a sewage sludge inoculum to assess the anaerobic biodegradability of several aliphatic polyester materials. Interestingly, while PHB/PHV biodegraded under these conditions, PCL did not. These observations may indicate that the extracellular depolymerase enzymes needed to initially attack PCL are not present in anaerobic sewage sludge. The American Society for Testing & Materials has established a similar test method using sewage sludge inoculum, designated D5210-92 (ASTM, 1992c). An alternative test method, recently proposed to the ASTM D.20.96 subcommittee on environmentally degradable plastics, involves determining the anaerobic biodegradability of test materials in solid waste anaerobic digestion conditions.

This methodology also measures CO_2 and CH_4 production but involves higher operating temperatures (52°C) and exposes the test materials to a suite of microorganisms associated with solid waste digestion. However, in the same manner that high biomass and high carbon conditions make aerobic compost and soil biodegradation tests difficult to design, it is difficult to measure the anaerobic biodegradability of a test material in a high solids anaerobic matrix. One solution is to design experiments that make use of specific detection methods such as the use of ^{14}C-labeled polymeric materials (see below).

6.5 CONFIRMATION LEVEL I
(LABORATORY- AND PILOT-SCALE TESTS)

Screening-level evaluations of materials do not provide definitive evidence of biodegradation (Kaplan et al., 1979). The possibility of overestimation of biodegradation potential exists if the "priming effect" occurs (Sharabi and Bartha, 1993), if material transformations are due to microbial attack on additives (ASTM, 1990) rather than mineralization of a polymeric component of a material, or if a material is exposed to microbial cultures that are not representative of the environment in which the material will be disposed. Alternatively, the presence of pro-oxidants or starches in a material may facilitate major physical changes (i.e., disintegration) that could be misinterpreted as evidence of complete biodegradation (Cole, 1990). Weight loss and tensile strength changes may be due to partial hydrolysis or abiotic hydrolysis caused by interaction of the polymer with the medium. Thus, more definitive biodegradation tests simulating the environment in which the polymer will ultimately reside are required to determine the practical biodegradability of polymeric materials in MSW waste treatment and terrestrial environments.

6.5.1 Isotopically Labeled Materials

The practical biodegradability of a polymeric material can be demonstrated by documenting complete mineralization of the polymer to CO_2 or CH_4 or as assimilation of decomposition residues into microbial biomass or stable humic substances under realistic environmental conditions. Definitive proof of biodegradation, or the lack thereof, may be provided through the use of isotopically labeled samples. A radioisotope of carbon (^{14}C) has found many applications in polymer research, especially in the design of experiments to run at realistic test substance concentrations (<1%) or for long time periods (one or more years). These tests have the advantage of being sensitive to small amounts of [14]CO_2 or [14]CH_4 evolution when the rate of mineralization is very slow, as was the case in a 10-year study of ^{14}C-polyethylene biodegradation in soil (Albertsson and Karlsson, 1988). Likewise, Kaplan et al. (1979) and Cook et al. (1994) found ^{14}C-labeled polymeric materials to be useful in experiments where background levels of respiration were high. They incubated ^{14}C-polystyrene in biologically active sludge and manure, and then trapped and analyzed gases generated during an 11-week incubation period. Total polystyrene decomposition was only 0.57% over 11 weeks. Degradation of such low magnitude would be difficult to detect without sensitive analytical techniques using ^{14}C-labeled polymeric materials.

Biodegradation tests conducted at the confirmation-level are designed to simulate actual environmental systems. Test conditions that mimic a compost or soil environment involve high biomass levels in a solid matrix and low test substance concentrations. Using PCL as a model substrate, Pettigrew et al. (1995) demonstrated that misleading or false negative results from screening level tests can be avoided using ^{14}C-labeled test materials and methods that simulate compost conditions. In a similar manner, Komarek et al. (1993) noted that ^{14}C-labeled polymers (cellulose acetate and cellulose propionate) can be used in experiments conducted under realistic and relevant conditions to facilitate the extrapolation of laboratory results to the environment. Scheu et al. (1993) noted that a radiolabel can be traced through incorporation into microbial biomass and used as additional evidence of microbial attack. In addition, Albertsson (1978) noted that sterilized control systems can be run alongside test systems to distinguish between abiotic and biotic transformation. However, complete sterilization of solid matrices such as soil is extremely difficult to achieve and maintain over time. Sterilized samples must be monitored carefully for the re-emergence of microbial populations, particularly spore-forming bacteria.

Researchers have been interested in assessing the biodegradability of synthetic polymeric materials in an anaerobic environment, given that the majority of MSW is disposed into landfills (USEPA, 1994). For example, Stegmann et al. (1993) and Pohland et al. (1993) assessed the fate of absorbent gelling material (AGM), a cross-linked polyacrylate used in disposable hygiene products, simulated landfill reactors (lysimeters) containing a mixture of solid waste, and hygiene products. In these studies, ^{14}C-labeled AGM was used to facilitate fate analyses in the solid matrix, leachate, and gas phases and to facilitate mass balance determinations. Both studies indicated that only minor degradation of the AGM occurred in the lysimeters, with less than 1% converted to CO_2 and CH_4 and from 2 to 4% appearing in leachate during 400- to 1700-day tests.

In addition to allowing biodegradation experiments to be run under realistic conditions, isotopically labeled samples also facilitate detailed investigation of polymer biodegradation mechanisms and degradation pathways. For example, Pitt and Schindler (1983) studied the *in vivo* biodegradability of ^{14}C-PCL and its metabolites by measuring the radioactivity that accumulated in the urine, feces, expired air, and implant site (subcutaneous) of rats. These authors demonstrated that weight loss occurred only when random chain hydrolysis produced oligomeric species small enough to diffuse from the bulk polymer. When undergoing ester hydrolysis, PCL produces the intermediate hexanedioic acid (adipic acid). Rusoff et al. (1960) conducted a study in which ^{14}C-adipic acid was fed to rats. Use of the radiotracer allowed the following adipic acid metabolites to be isolated and identified: urea, glutamic acid, lactic acid, beta-ketoadipic acid, and citric acid. The presence of radiolabeled beta-ketoadipic acid suggested that adipic acid was metabolized initially by the beta-oxidation pathway. In addition, identification of citric acid and glutamic acid metabolites suggested subsequent degradation via the tricarboxylic acid (Krebs) pathway.

Despite the analytical advantages that ^{14}C-labeled polymeric materials provide, the interpretations and conclusions of these studies are only as good as the purity and quality of the test material. Due to cost and nuclear regulatory constraints, the synthesis of a ^{14}C-labeled polymeric material must be done with specialized equipment and at a small scale. These conditions can make it difficult to generate a labeled material that is truly representative of a commercial polymer. In addition, the synthesis and characterization of ^{14}C-labeled polymeric materials must ensure that the position of the ^{14}C label is understood and that the potential formation of ^{14}C-labeled, low molecular weight oligomers and impurities is addressed.

An alternative approach that uses isotopically labeled samples is based on the analysis of the stable carbon isotope (^{13}C). When complex heteropolymers contain polymer components that have different ^{13}C concentrations, either through natural abundance or chemical synthesis, the biodegradability of the individual components can be analyzed using standard methods of mass spectrometry. Sykes et al. (1994) presented preliminary results on starch-synthetic polymer blends and demonstrated that this approach can provide a quantitative assessment of the degradation of each component. However, these authors also noted that further work is necessary to confirm fractionation effects of processing, chemical degradation, and biodegradation.

6.5.2 Mesocosm Methods

The screening-level and confirmatory-level tests described above have the common limitation that they use *in vitro* batch systems which ultimately encounter nutrient limitations and lack mechanisms for product removal. While such closed batch systems facilitate mass balance determinations, the conditions used in these tests do not mimic the open conditions found in some waste treatment systems and the environment that may affect the realistic biodegradability of a test material. In addition, *in vitro* batch systems often fail to simulate all the biotic and abiotic factors that interact to facilitate the biodegradation of polymeric materials in natural environments. For example, MSW compost systems rely on a significant amount of mechanical energy to degrade polymeric materials. Such conditions are lacking in most *in vitro* batch tests. Alternative confirmatory-level biodegradation tests using mesocosm designs that simulate "full-scale" systems are required to avoid these limitations. While larger than microcosms, mesocosm simulations usually are conducted on reduced physical and temporal scales relative to field scale to facilitate accuracy and precision.

Mesocosm systems have been developed to simulate the physical, chemical, thermal, and biological features of MSW compost or leaf compost environments. For example, one version of such a test system is a rotating drum that simulates the initial mixing of MSW for composting (Smith et al., 1992; Schwab et al., 1994). Continuous mixing in this type of apparatus helps to ensure uniform exposure of test material to compost microorganisms and to provide mechanical agitation and adequate aeration. Mesocosm systems have also been designed to create an environment in which synthetic polymers are exposed to a diverse microbial community actively degrading an organic feedstock. Krupp and Jewell (1992) monitored the anaerobic and aerobic biodegradability of plastic films using large cylindrical tanks that were fed with specific mixtures of test materials, sorghum (to provide nutrients and an alternative carbon source), and cellulose at a loading rate of 1.42 g of volatile solids per kilogram of total reactor contents per day. The anaerobic reactor was operated for 350 days at 58°C with 40 kg of total reactor contents. The aerobic reactor was operated for 185 days at 37°C with 30 kg of total reactor contents. Using mass loss and biological oxygen demand as biodegradation endpoints, Krupp and Jewell (1992) noted that only PHB/PHV films showed evidence of substantial degradation and that polyethylene plastic films containing starch showed no evidence that anything other than the starch was biodegraded.

In a review of trends in municipal solid waste management, Narayan (1993) discussed the need to equip pilot-scale (mesocosm) compost reactors with monitoring devices to help researchers ensure the system functions realistically and consistently. In particular, CO_2 production or O_2 consumption should be monitored as an indication of overall (test material and organic feedstock) mineralization over time. Narayan (1993) also noted that reactors should be equipped with a moisture probe to ensure that the composting feedstock becomes neither too dry (which would slow microbial decomposition) nor too wet (which would create an anaerobic condition). An initial moisture content of 60% was used by Narayan; however, a general rule of 50 to 70% moisture is appropriate, depending upon the feedstock and system design. Formulation of the compost feedstock also should be tailored to fit the research objectives. Smith et al. (1992) and Schwab et al. (1994) used a synthetic waste matrix that was similar to actual U.S. solid waste in terms of percent protein, simple carbohydrates, lipids, cellulosics, and inert materials. These investigators concluded that the simulated solid wastes underwent physical, chemical, and microbiological changes similar to those that occur in full-scale composting systems.

6.5.3 Residue and Degradation Product Analysis

Because synthetic polymeric materials are rarely completely mineralized to CO_2 or CH_4 and because test materials often degrade to the point at which they cannot be visually distinguished from other components in the test matrix, it is important to consider what is "left over" following a biodegradation test. Additional test methods are required for determination of the residues and degradation products of a test material. Because it is difficult to subtract background interferences, these methods generally provide limited quantitative information about the biodegradation of a material and typically are used in combination with other quantitative test methods.

Hydrolysis of high molecular weight (>5000) polymers to yield lower molecular weight segments is a typical first step in the process of polymer biodegradation. For many polymeric materials, no significant mass loss is evident until polymer chain length is reduced abiotically or by exogenous enzymes to a given threshold. Fields et al. (1974) found that the biodegradation of PCL was directly proportional to the content of low molecular weight segments in screening-level tests with *Pullularia pullulans*. Potts et al. (1973) reported similar observations in mineralization studies on straight chain hydrocarbons. At molecular weights below 451 (corresponding to 32-carbon chains), test compounds supported abundant growth of *Aspergillis flavus, Chaetomium globosum,* and *Penicillium funiculosum*. Above this molecular weight, no fungal growth was observed on the straight chain alkanes. The authors speculated that enzymes typically responsible for beta-oxidation are unable to complex with chain ends in the high molecular weight materials due to stearic hindrance.

As degradation occurs, changes in molecular weight distribution can be monitored by the use of gel permeation chromatography (GPC). In this technique, materials are extracted with an organic solvent or recovered from a test system, then filtered and injected onto a liquid chromatography column capable of resolving a standard range of molecular weights (2000 up to 100,000). Molecular weights of eluted fractions are calculated relative to standard polymers such as polystyrene (Gu et al., 1993) or polyisoprene (Tsuchii et al., 1985). Provided a linear relationship exists between the detector response and polymer concentration, changes in peak height or area and the appearance of new peaks following exposure to microorganisms can be used as an indication of biodegradation. This type of residual analysis was demonstrated by Benedict and coworkers (1983a,b) in studies of PCL biodegradation (Figure 5). Microbial cultures grown in the absence of polymer can be processed as controls to account for any microbial material solubilized during processing of polymer. Also, polymers of varying chain length can be added to microbial cultures just prior to harvest and processed to account for sorption of polymer to microbial biomass.

Use of GPC techniques with complex soil or compost matrices requires careful sample extraction and filtration to remove high molecular weight components of soil (e.g., humic materials) that may interfere with the detection of synthetic polymers. In addition, the analysis of molecular weight distribution of some polymeric materials does present difficulties due to differential adsorption phenomena, particularly inter- and intramolecular solute interactions with confounded molecular size distributions (Rausa et al., 1991). In studies

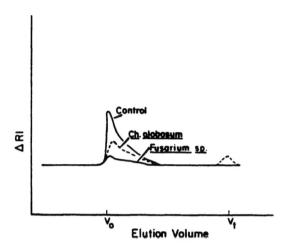

FIGURE 5   Gel permeation chromatogram of polycaprolactone 700 (mean mol. wt. 35,000) control and reduced high molecular weight peak and low molecular weight materials produced by *Chaetomium globosum* (ATCC 6205). Also shown is the reduced high molecular weight peak for a *Fusarium* sp. RI represents the refractive index detected using a R-4 Differential Refractometer. V_o represents the upper exclusion limit at which polymers eluted. V_t represents the lower exclusion limit at which low molecular weight residual materials eluted. (From Benedict, C.V., W.J. Cook, P. Jarrett, J.A. Cameron, S.J. Huang, and J.P Bell. 1983b. *J. Appl. Polym. Sci.* 28:331. With permission from John Wiley & Sons, New York.)

with natural humic materials, they stressed the importance of using short, high performance columns to ensure reproducible chromatographic peaks.

Fourier transform infrared (FTIR) spectra can be used for the semiquantitative analysis of the chemical composition of complex organic materials (Baes and Bloom, 1989). The analysis is based on the absorption of different wavelengths of infrared radiation by bonded atoms in a sample, with frequency of absorption being a function of vibrational characteristics of the atoms and their bond strengths and masses (Bloom and Leenheer, 1989). Based on surveys of fulvic and humic acids recovered from soils and composts (Baes and Bloom, 1989; Inbar et al., 1989), various absorption peaks can be assigned to numerous C–C, C–O, C—H, C=C, and C–peptide bonds, as well as aliphatic and phenolic OH groups.

Fourier transform infrared spectroscopy is of high diagnostic value, particularly for identification of molecular components and functional groups. FTIR spectroscopy has been used to study the degradation of synthetic polymers at the molecular level. For example, Greizerstein et al. (1993) analyzed the spectral regions of polyethylene-based materials between 4000 and 400 cm^{-1} at 2-wave number resolution. The peak and area of individual absorbance bands of fresh and fungal-decayed samples were compared to discern changes in chemical functional groups and to indicate oxidative reactions within the polymer. The results presented in Figure 6 indicate that the ratio of the ester

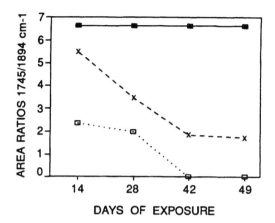

FIGURE 6 Ratio of Fourier transform infrared (FTIR) absorbence at 1745 cm^{-1} and 1894 cm^{-1} bands in starch-modified polythylene bags. Nonexposed control (——), buried (·····), and surface-exposed (----) bags are represented. The carbonyl band at 1745 cm^{-1} is assigned to an ester additive. The band at 1894 cm^{-1} is assigned to the crystaline phase of polyethylene films. (From Greizerstein, H.B., J.A. Syracruse, and P.J. Kostyniak. 1993. *Poly. Degrad. Stabil.* 39:254. With permission from Elsevier Science Publishers Ltd., Kidlington, UK.)

additive band (1745 cm^{-1}) and crystalline polyethylene band (1894 cm^{-1}) decreased with time in starch/polyethylene films exposed to compost. The loss of functional groups was demonstrated using FTIR on samples of lignin-styrene graft copolymers exposed to pure fungal cultures (Milstein et al., 1994). The absorbance spectrum of the exposed sample was corrected for both the weight percent of fungi present and the absorbance spectrum of the myce-lial mat. This was particularly useful because the fungus was intimately bound or physically integrated into the sample. Otherwise, Janshekar et al. (1981) advised that analyses of polymers must be free of microorganisms so that extraction of cell matter does not influence banding patterns.

The use of FTIR spectroscopy has two predominant limitations. First, samples usually are prepared using the potassium bromide (KBr) pressed pellet method (Schnitzer, 1971). Potassium bromide is hygroscopic; thus, it is diffi-cult to prepare pellets without moisture which interferes with some absorption bands. Also, the high temperature and pressure required for pellet formation may alter physical and chemical features of the polymer sample. Baes and Bloom (1989) suggested the use of diffuse reflectance infrared Fourier trans-form (DRIFT) to overcome these limitations. Diffuse reflectance measure-ments do not require the KBr pressed pellet methodology, and vacuum drying can be used in place of high temperatures. An absorption spectrum is obtained through a wave number range of 4000 to 400 cm^{-1} and then plotted in "Kubelka-Munk" (K-M) units which are a function of absorption and reflection coefficients. The theoretical physics for DRIFT have been reviewed by Childers and Palmer (1986).

Processes of polymer degradation and stabilization in natural environments yield products that are chemically complex and difficult to fractionate. Chen et al. (1989) cautioned that alkali extraction procedures typically used for isolating humic substances may extract only a portion of the carbon from the system. These researchers suggested that the need to profile the total organic content of an environmental matrix, including degradation products of test materials, is best met by methods that use intact samples with a minimum of chemical treatment. Nuclear magnetic resonance (NMR) spectroscopy is one such method. It has been applied to study the microbial degradation of compost organic matter (Inbar et al., 1989), cellulose acetate (Gu et al., 1993), lignin (Robert and Chen, 1989; Martinez et al., 1991) and synthetic polymers (Milstein et al., 1994). One common approach to NMR application on solid samples is "cross polarization" and "magnetic-angle spinning" (CPMAS). Each signal, defined as a chemical shift, is assigned to a carbon atom at a particular location in a unique aromatic or linear chain or, otherwise, a component of a particular ester or ether linkage. For analysis of wood residues, CPMAS NMR has been used to measure proportions of total carbon in carboxyl, aromatic carbohydrate, or aliphatic groups (Chen et al., 1989; Martinez et al., 1991). By monitoring changes with fresh and exposed lignin samples, Robert and Chen (1989) were able to determine the extent of oxidative cleavage of both side chains and aromatic rings caused by the white rot fungus *Phanerochaete chrysosporium*. NMR methods have been most widely applied to studies of soil organic matter. It is nondestructive, requires small sample weights, involves limited manipulations, and is of particular value in quantitative analysis of structural components (Stevenson, 1994).

The mechanism of microbial attack on a complex material is of great interest to polymer scientists; however, obtaining direct analytical proof of these mechanisms is often difficult. Some analytical methods involve extensive depolymerization of the parent material using chemical procedures that minimize side reactions to produce monomers and oligomers that can be readily analyzed. For lignin, one approach has been the analysis of components in fresh and exposed samples using cupric oxide (CuO) alkaline degradation (Hedges et al., 1988; Martinez et al., 1991). Products from the CuO degradation are separated and quantified using gas chromatography. Using this technique, it is possible to determine the relative proportion (in μmol per gram of sample) of such lignin constituents as guaiacyl, syringyl, cinnamyl, and vanillyl phenols (Martinez et al., 1991; Alberts et al., 1992). Comparing molar ratios of different constituents indicates the extent of oxidative degradation and the degree to which different constituents have been degraded. For example, Hedges et al. (1988) used the technique to demonstrate that white rot fungi indiscriminately decompose both polysaccharides and lignin in wood samples, whereas brown rot fungi preferentially act on polysaccharides, relative to lignin.

Chemical structural analysis of complex materials can be approached also through the use of Curie-point pyrolysis-gas chromatography mass spectrometry (Py-GC-MS). The method involves the controlled combustion of a sample, followed by collection and chromatographic separation of pyrolysis products (Garcia et al., 1992). Peaks on the resultant chromatogram correspond to unique sample constituents. Comparisons of fresh and exposed materials indicate which sample constituents have been degraded and also indicate the amount and type of by-products from microbial attack (Saiz-Jimenez and De Leeuw, 1984). An advantage of Py-GC-MS is that samples are assayed without extraction. This avoids the use of chemical degradation techniques, which may be tedious and prone to affect changes or losses of key sample components.

6.6 CONFIRMATION LEVEL II (FIELD- OR FULL-SCALE TESTS)

Screening and Confirmation Level I test methods used to assess the biodegradation of polymeric materials are short-term tests run under narrowly defined environmental conditions. Although these tests may be conducted under conditions that mimic the environment, the biodegradability of a material may require verification in field- or full-scale tests where there is a greater diversity of biotic communities and a range of moisture availability, nutritional status, pH, oxygen availability, and temperature.

6.6.1 Biodegradation in Solid Waste Treatment and Disposal

As described in Chapter 2, polymer biodegradability in landfills is extremely variable depending on the biotic and abiotic factors at a given site. Extreme heterogeneity between and within landfill sites and the practical problems of accessing buried samples (Suflita et al., 1992) have resulted in few investigations of *in situ* landfill biodegradation that focus on assessing the fate of synthetic polymers. Consequently, the biodegradation processes for synthetic polymers in landfills are not well known. However, Breslin (1993) followed the rate and extent of biodegradation of starch-plastic composites buried for up to 2 years in an MSW landfill. This investigator monitored changes in the starch content, tensile properties, and weight loss of polymer samples recovered from the landfill. He found that while the starch content decreased approximately 25 to 33%, the starch-plastic composites did not fragment during the study.

Published research on the *in situ* biodegradability of synthetic polymers in MSW treatment systems is limited. A notable exception involves research that was done on the fate of several plastic films in a municipal leaf-composting facility (Gilmore et al., 1992). These investigators monitored changes in molecular weight, tensile properties, and weight loss of polymer samples recovered from the compost system for 6 months. While very little change was noted for starch-polyolefin plastics, significant deterioration and weight loss of polyhydroxyalkanoate plastics was observed. Using the clear-zone screening assay previously described, the investigators also screened microbial isolates obtained from the compost matrix and found that only the fungal isolates possessed PHB/PHV depolymerase activity.

6.6.2 Biodegradation in Terrestrial Ecosystems

Field-level terrestrial biodegradation tests generally involve burial of plastic materials in soil for an extended period of time. The time frame is dictated by the rate of physical and chemical disintegration of the material. Soil pH, moisture availability, and fertility are key factors that investigators should consider when choosing a soil matrix (Dibble and Bartha, 1979; Yabannavar and Bartha, 1993). For example, Genouw et al. (1994) observed increased biodegradation of organic materials in soil amended with mineral nutrients or compost as an organic carbon supplement. While this work pertained to oil-contaminated sludge, the technique is applicable to the land application of composts containing synthetic polymers. Field tests that have assessed *in situ* soil biodegradation include rubber (Williams, 1982) and lignin (Laishram and Yadava, 1988), as well as synthetic materials such as PHB/PHV (Barak et al., 1991), PCL (Potts et al., 1973), and polyethylene (Karlsson et al., 1988). Besides soil, freshwater sediments have been used to demonstrate the *in situ* biodegradation of PHB/PHV materials (Brandl and Puchner, 1992).

The presence of saprophytic invertebrates may be an advantage with field tests that is difficult to achieve in laboratory systems. By a comminution effect on complex organic residues in soils or composts, earthworms increase the surface area exposed to microorganisms and facilitate inoculation of the polymer surface (Saiz-Jimenez et al., 1989). Fungi and bacteria typically are attached to solid surfaces, and there is little facile movement from attachment sites to adjacent polymer surfaces (Cole, 1990). Curry and Byrne (1992) reported that the decomposition rate of straw accessible to earthworms in soil increased by up to 47% over 10 months relative to straw from which earthworms were excluded. Similar positive effects may be observed with synthetic polymers, but few data are available in the literature to confirm this.

One limitation of field studies is that it is impossible to maintain abiotic controls over long periods. While control soil plots may be fumigated with methyl bromide or chloropicrin (Lamar and Dietrich, 1990), sterility of controls cannot be maintained for long-term studies. Chemical and physical changes seen in materials at various stages of decay in long-term tests may be attributed to solid waste or soil biota or to abiotic effects of sunlight, oxygen, pH, water or even freeze-thaw cycles. In addition, the heterogeneous nature of soils and compost and the subsequent variation that occurs between subsamples limit the accuracy of any measurement, including biodegradation assessments. For example, Mullins and Hutchinson (1982) found that the variability between subsamples of soils was shown to depend not only on the sampling procedure adopted, but also on the physical properties of the sample itself. On a practical level, *in situ* methods require large amounts of test materials and are limited by the inability to measure mineralization quantitatively. Nonetheless, by thoroughly mixing a material into MSW or soil and by having adequate methods to recover materials, field tests can provide evidence of *in situ* biodegradation and the extent to which it occurs.

6.7 STANDARDIZATION OF TEST METHODS

The generally recognized need for the standardization of polymer biodegradation test methods has resulted in the creation of working groups such as the Degradable Polymeric Materials Advisory Committee in the American Society for Testing and Materials Institute for Standards Research (ASTM/ISR) and the Compostability Criteria Technical Committee in the European Organic Reclamation and Composting Association (ORCA). Both of these groups were chartered to provide the scientific substantiation of disposability statements for biodegradable polymeric materials in solid waste treatment systems. One of the goals of these organizations is to identify standard test methods that can be used to assess the environmental fate of plastic materials in solid waste treatment and terrestrial environments. To accomplish this goal, the ASTM/ISR has initiated a research program to address the limitations, performance, and accuracy of test methods used at all levels of the tiered testing approach described in this chapter. This work is not yet complete and specific recommendations have not been made. However, it is likely that careful consideration of test conditions, elucidation of the correlations between test methods, and the establishment of appropriate biodegradation benchmarks will be required for the standardization of polymer biodegradation test methods.

6.7.1 Test Conditions

Polymer biodegradation test results can be affected by the physical and chemical properties of polymeric samples as well as numerous environmental variables (Table 2). Another important factor to consider is the form of the polymeric material added to the biodegradation test. In particular, the particle size and surface area of the polymeric sample have been shown to affect the biodegradability of polymeric materials (Yakabe and Tadakoro, 1993; Pettigrew et al., 1995). Consideration of the concentration of the test material used will also be important for determining the validity and cross comparison of data generated from various biodegradation tests. Some investigators have noted that the presence of large amounts of polymer may lead to a high concentration of degradation products that could substantially alter chemical/physical properties or otherwise elicit a toxic effect on microbes (Karlsson et al., 1988; Pettigrew et al., 1995). As discussed previously, the various screening and confirmatory level tests use specific test conditions that limit how well the tests predict what happens in the natural environment.

6.7.2 Comparisons Among Tests

As summarized in Table 4, each of the test methods described in the tiered testing approach have strengths and limitations that will affect the quality and usefulness of the data that are generated. Screening level test methods have been used most often to assess the biodegradability of synthetic polymers; however, these test methods must be evaluated very carefully since the test

**TABLE 4   Summary of the Strengths and Limitations of Biodegradation Data
Generated at Different Levels of a Tiered Testing Approach**

Source of Data	Strengths	Limitations
Screening level tests	Relatively rapid test methods facilitate testing of wide range of materials. Numerous data sets are available. Test designs allow multiple observations and replication.	Tests are not environmentally relevant and data are difficult to extrapolate to the environment. Often uses indirect assays. False negatives are common.
Confirmatory level tests	Biodegradation data are environmentally relevant. Direct assays use sensitive techniques. Tests facilitate control of variables.	Tests require specialized equipment and techniques. Relatively few data sets are available.
Field tests	Tests provide confirmation of predicted environmental fate and rate measurements under natural conditions.	Confounding factors that impact biodegradation cannot be controlled. Removal may be due to abiotic factors. Slow rates of biodegradtion result in lengthy test duration. Data base is extremely limited.

conditions are not environmentally relevant, and numerous false-negative and false-positive results have been documented. Thus, such tests can provide only preliminary conclusions. Yabannavar and Bartha (1994) recently noted the paucity of published reports that compare the performance and accuracy of the various screening level test methods in a statistically controlled manner. These authors found that tensile strength decline and CO_2 evolution do not necessarily correlate with molecular weight decline as determined by GPC analysis. Confirmation Level I and II test methods, wherein the practical biodegradability of a material is determined, provide the most relevant information. Unfortunately, confirmation and field level data are very limited for synthetic polymeric materials, and there have been few attempts to compare these data with screening level data. Using PCL as an example of a synthetic polymer that is generally regarded as biodegradable and compostable, Fields et al. (1974) found good correlation between Screening and Confirmation Level II test methods. In contrast, Pettigrew et al. (1995) found that the rate and extent of PCL biodegradation predicted by screening level test methods were much lower than the rate and extent measured in Confirmation Level II test methods using ^{14}C-labeled PCL. These authors noted that, while screening level tests provide valuable information about the inherent biodegradability of a polymer, caution must be used in predicting the environmental fate of polymeric materials on the basis of screening level tests. In most cases, supplemental chemical analyses or confirmatory biodegradation tests must be performed to confirm synthetic polymer biodegradation and to ensure that no residues or intermediates persist. The use of polymer standards in studies of synthetic polymer biodegradation can also facilitate the comparison of data generated from different tests. The rates of synthetic polymer biodegradation can be normalized with respect to a standard polymer included in all tests (e.g., thin layer chromatography or filter paper cellulose), thereby facilitating cross comparisons between tests.

6.7.3 Biodegradation Benchmarks

The biodegradation of new chemicals with discrete molecular weights has been referenced historically against pure homogeneous compounds. In contrast, many new synthetic polymeric materials are complex assemblages of heterogeneous compounds with a range of biodegradabilities. One approach to assessing the biodegradability of complex materials involves benchmarking synthetic polymer biodegradation test results against results from natural materials that are generally regarded as biodegradable but are equally complex and heterogeneous. For example, lignocellulose has been studied widely in terms of the mechanisms and kinetics of biodegradation, and it is one of the most abundant natural polymers on earth (Crawford, 1981). Lignin is chemically complex, being built by enzymatic and autooxidative reactions involving a variety of different monomers (Stevenson, 1994). Contained within lignin is a variety of bonds, many of which are not readily hydrolyzable. Thus, lignocellulose possesses general resistance to rapid decomposition. These features make lignin and lignocellulose good benchmark compounds for the study of synthetic polymer biodegradation. The results presented in Figure 4 demonstrate that materials composed of complex assemblages of polymers will degrade much differently than a homopolymer such as cellulose. These data illustrate the relative biodegradation of materials containing varying amounts of lignocellulose and suggest that the time scales for the biodegradation of complex assemblages of polymers is from months to years rather than hours to days. Fortunately, the microbial decomposition of lignocellulose can be studied under realistic environmental conditions using ^{14}C-(cellulose)-lignocellulose synthesized by the methods and procedures reviewed by Crawford (1981).

6.8 CONCLUSIONS

6.8.1 Putting Biodegradation Data to Use

The use and release of synthetic polymeric materials in the environment has led to the development of test methods that help predict their fate in MSW treatment and disposal systems and in terrestrial environments. Determination of the rate and extent of biodegradation are the key data necessary to assess environmental fate. These data are also used to estimate environmental concentrations and the extent to which a material might accumulate over time, especially under conditions where the polymer is applied repeatedly to a soil as a direct amendment or as a component of compost. Accumulation in soil is heavily dependent upon the biodegradation rate for a material. Under the most conservative circumstances, no biodegradation can be assumed, and the material will accumulate linearly as a function of the application frequency until the application ceases. However, if the material does biodegrade, these conservative calculations will dramatically overestimate the "real world" concentrations. Under these circumstances, it is more appropriate to predict accumulation potential using mathematical models that consider biodegradation. Historically, the biodegradation of low molecular weight chemicals in aquatic environments has often been modeled by first order kinetics (Shimp et al., 1990). However, there are numerous alternative models that can be used to describe biodegradation data and these may prove useful in modeling the biodegradation of heterogeneous, high molecular weight materials (Battersby, 1990).

This chapter has focused only on methods relating to the environmental fate and estimating the ultimate exposure of synthetic polymers disposed with solid waste. The ultimate assessment of the environmental risks and benefits associated with using biodegradable synthetic polymers will require additional information. A holistic environmental evaluation requires an understanding of the environmental impacts associated with the acquisition of raw materials, manufacturing process, product use/reuse, and disposal. In addition to understanding the fate of synthetic plastics, assessing the potential environmental effects of these materials and their degradation products is also required (Shimp, 1993). An array of terrestrial effects tests already exists and should facilitate the development of a research program that is focused on bridging polymer fate and effects testing. Thus, while the specific circumstances of MSW disposal require different consideration than the better established area of aquatic safety assessment, the conceptual framework and many experimental tools are already available for conducting environmental safety assessments. Direct toxicity studies can be conducted in laboratory, microcosm, and field assays. In addition, indirect toxicity assays are available to measure potential effects that are the result of alterations in the physical or chemical characteristics of polymer-amended soils. Other questions that should be asked include: (1) are the materials or their degradation products mobile in aquatic environments, and (2) to what extent can the material or its degradation products bioaccumulate in plants or animals? Once these questions have been answered and environmental effects data are available, all the fate and effects data can be integrated into an environmental safety assessment for a polymeric material.

6.8.2 Current Status and Research Opportunities

The best opportunities for the utilization of biodegradable polymers to address solid waste issues are found in emerging composting and managed landfill operations. In North America, the aerobic biodegradation of organic materials is being promoted as a means of recovering organic waste (Composting Council, 1994; USEPA, 1994). Thus, it will be increasingly important to have approaches that accurately predict the fate of polymers in compost environments. Assessing the fate of polymeric materials associated with compost that is subsequently land-applied also represents a promising research area. Likewise, a better understanding of the anaerobic biodegradability of organic materials in landfills is needed. In other geographies (e.g., Europe), additional emphasis is placed on understanding the anaerobic biodegradability of polymeric materials that are disposed via high solids anaerobic digesters ("anaerobic biogasification").

The recent interest in the development of biodegradable polymers has resulted in an increase in scientific research and publications in the field, the creation of new scientific societies (Biodegradable Polymer Society in Japan and the Bio/Environmentally Degradable Polymer Society in North America), and numerous symposia and conferences. The specific advantages and limitations of existing biodegradation test methods, as well as the introduction of new methodologies, have received much attention in these scientific forums and are expected to remain a high priority in the future. In addition, the ASTM/ISR and ORCA organizations continue to develop useful definitions for biodegradable polymeric materials and guidelines for evaluating new polymers. However, much work remains regarding the standardization and correlation of biodegradation test methods and extrapolation of data to describe the ultimate fate of synthetic polymers in the environment. In particular, an enhanced quantitative understanding of how various physical, chemical, and biological factors influence the rate and extent of synthetic polymer biodegradation is badly needed.

ACKNOWLEDGMENTS

The authors thank D.J. Kain, K. Mesuere, and R.J. Shimp for reviewing this manuscript and providing stimulating discussions about this work. The authors also thank T.W. Federle and S.D. Goodwin for sharing unpublished material.

REFERENCES

Abe, H., I. Matsubara, Y. Doi, Y. Hori, and A. Yamaguchi. 1994. Physical properties and enzymatic degradability of poly(3-hydroxybutyrate) stereoisomers with different stereoregularities. *Macromolecules* 27:6018–6025.

Adney, W.S., C.J. Rivard, K. Grohmann, and M.E. Himmel. 1989. Detection of extracellular hydrolytic enzymes in the anaerobic digestion of municipal solid waste. *Biotech. Appl. Biochem.* 11:387–400.

Alberts, J.J., Z. Filip, M.T. Price, J.I. Hedges, and T.R. Jacobsen. 1992. CuO-oxidation products, acid hydrolyzable monosaccharides and amino acids of humic substances occurring in a salt marsh estuary. *Org. Geochem.* 18:171–180.

Albertsson, A.-C. 1978. Biodegradation of synthetic polymers. II. A limited microbial conversion of ^{14}C in polyethylene to $^{14}CO_2$ by some soil fungi. *J. Appl. Polym. Sci.* 22:3419–3433.

Albertsson, A.-C. and S. Karlsson, 1988. The three stages in degradation of polymers — polyethylene as a model substance. *J. Appl. Polym. Sci.* 35:1289–1302.

Alexander, M. 1973. Nondegradable and other recalcitrant molecules. *Biotech. Bioeng.* 15:611–647.

APHA. 1989. *Standard Methods for the Examination of Waters and Waste-Waters,* 20th ed., American Public Health Association, Washington, D.C.

ASTM. 1980. Standard practice for determining resistence of synthetic polymeric materials to bacteria, Designation G22-76. American Society for Testing and Materials, Philadelphia, PA.

ASTM. 1986. Standard method of accelerated laboratory test of natural decay resistance of woods, Designation D2017. American Society for Testing and Materials, Philadelphia, PA.

ASTM. 1990. Standard practice for determining resistence of synthetic polymeric materials to fungi, Designation G21. American Society for Testing and Materials, Philadelphia, PA.

ASTM. 1991. Standard test method for mold growth resistance of blue stock (leather), Designation D4576. American Society for Testing and Materials, Philadelphia, PA.

ASTM. 1992a. Standard test method for determining the aerobic biodegradation of plastic materials in the presence of municipal sewage sludge, Designation D 5209–92. American Society for Testing and Materials, Philadelphia, PA.

ASTM. 1992b. Standard test method for determining the aerobic biodegradation of plastic materials under controlled composting conditions, Designation D 5338–92. American Society for Testing and Materials, Philadelphia, PA.

ASTM. 1992c. Standard test method for determining the anaerobic biodegradation of plastic materials in the presence of municipal sewage sludge, Designation D 5210-92. American Society for Testing and Materials, Philadelphia, PA.

ASTM. 1993. Standard test method for determining the aerobic biodegradation of plastic materials in an activated-sludge wastewater-treatment system, Designation D 5271-93. American Society for Testing and Materials, Philadelphia, PA.

Baes, A.U. and P.R. Bloom. 1989. Diffuse reflectance and transmission Fourier transform infrared (DRIFT) spectroscopy of humic and fulvic acids. *Soil Sci. Soc. Am. J.* 53:695–700.

Barak, P., Y. Coquet, T.R. Halbach, and J.A.E. Molina. 1991. Biodegradability of polyhydroxybutryate(co-hydroxyvalerate) and starch-incorporated polyethylene plastic films in soils. *J. Environ. Qual.* 29:173–179.

Battersby, N.S. 1990. A review of biodegradation kinetics in the aquatic environment. *Chemosphere* 21:1243–1284.

Benedict, C.V., J.A. Cameron, and S.J. Huang. 1983a. Polycaprolactone degradation by mixed and pure cultures of bacteria and yeast. *J. Appl. Polym. Sci.* 28:335–342.

Benedict, C.V., W.J. Cook, P. Jarrett, J.A. Cameron, S.J. Huang, and J.P. Bell. 1983b. Fungal degradation of polycaprolactones. *J. Appl. Polym. Sci.* 28:327–334.

Beyea, J., L. DeChant, B. Jones, and M. Conditt. 1992. Composting plus recycling equals 70 percent diversion. *BioCycle* May:72–75.

Blondeau, R. 1989. Biodegradation of natural and synthetic humic acids by the white rot fungus *Phanerochaete chysosporium. Appl. Environ. Microbiol.* 55:1282–1285.

Bloom, P.R. and J.A. Leenheer. 1989. Vibrational, electronic and high-energy spectroscopic methods for characterizing humic substances, in M.H.B. Hayes, P. MacCarthy, R.L. Malcolm, and R.S. Swift (Eds.), *Humic Substances II,* John Wiley & Sons, New York, pp. 409–446.

Borel, M., A. Kergomard, and M.F. Renard. 1982. Degradation of natural rubber from *Fungi imperfecti. Agric. Biol. Chem.* 46:877–881.

Brandl, H., and P. Puchner. 1992. Biodegradation of plastic bottles made from "Biopol" in an aquatic ecosystem under *in situ* conditions. *Biodegradation* 2:237–243.

Breslin, V.T. 1993. Degradation of starch-plastic composites in a municipal solid waste landfill. *J. Environ. Polym. Degrad.* 1:127–141.

Burgess, R., and A.E. Darby. 1964. Two tests for the assessment of microbiological activity on plastics. *Br. Plast.* 37:32–37.

Burns, R.G. 1982. Enzyme activity in soil: location and possible role in microbial ecology. *Soil Biol. Biochem.* 14:423–427.

Chen, Y., Y. Inbar, Y. Hadar, and R.L. Malcolm. 1989. Chemical properties and solid state CPMAS ^{13}C-NMR of composted organic matter. *Sci. Total Environ.* 81/82:201–208.

Childers, J.W. and R.A Palmer. 1986. A comparison of photoacoustic and diffuse reflectance detection in FTIR spectrometry. *Am. Lab.* 18:22–38.

Cole, M.A. 1990. Constraints on decay of polysaccharide-plastic blends, in J.E. Glass and G. Swift (Eds.), *Agricultural and Synthetic Polymers: Biodegradation and Utilization,* American Chemical Society, Washington, D.C., pp. 76–95.

Composting Council. 1994. *Compost Facility Operating Guide: A Reference Guide for Composting Facility and Process Management,* The Composting Council, Alexandria, VA.

Cook, B.D., P.R. Bloom, and T.R. Halbach. 1994. A method for determining the ultimate fate of synthetic chemicals during composting. *Compost Sci. Util.* Winter 1994: 42–50.

Crawford, R.L. 1981. *Lignin Biodegradation and Transformation.* John Wiley & Sons, New York.

CSA. 1992. Test method for determining the anaerobic biodegradability of plastic materials, Designation CAN/CSA-Z218-M. Canadian Standards Association.

Curry, J.P. and D. Byrne. 1992. The role of earthworms in straw decomposition and nitrogen turnover in arable land in Ireland. *Soil Biol. Biochem.* 24:1409–1412.

Darby, R.T., and A.M. Kaplan. 1968. Fungal susceptibility of polyurethanes. *Appl. Microbiol.* 16:900–905.

Day, M., K. Shaw, and D. Cooney. 1994. Biodegradability: an assessment of commercial polymers according to the Canadian method for anaerobic conditions. *J. Environ. Polym. Degrad.* 2:121–127.

Dibble, J.T. and R. Bartha. 1979. The effect of environmental parameters on the biodegradation of oil sludge. *Appl. Environ. Microbiol.* 37:729–739.

Dimond, A.E., and J.G. Horsfall. 1943. Preventing the bacterial oxidation of rubber. *Science* 97:144–145.

Doi, Y. 1990. Biodegradation of microbial polyesters, in *Microbial Polyesters,* VCH Publishers, New York, pp. 135–152.

Federle, T.W. Unpublished material.

Fields, R.D., F. Rodriguez, and R.K. Finn. 1974. Microbial degradation of polyesters: Polycaprolactone degraded by *P. pullulans. J. Appl. Polym. Sci.* 18:3571–3579.

Fischer, W.K. 1971. Testing nonionic surfactants in the Closed Bottle test. *Tenside* 8:177–182, 182–188.

Garcia, C., T. Hernandez, F. Costa, B. Ceccanti, and M. Calcinai. 1992. A chemical-structural study of organic wastes and their humic acids during composting by means of pyrolysis-gas chromatography. *Sci. Total Environ.* 119:157–168.

Genouw, G., F. De Naeyer, P. Van Meenen, H. Van de Werf, W. De Nijs, and W. Verstraete. 1994. Degradation of oil sludge by land farming — a case study at the Ghent harbour. *Biodegradation* 5:37–46.

Gilmore, D.F., S. Antoun, R.W. Lenz, S. Goodwin, R. Austin, and R.C. Fuller. 1992. The fate of "biodegradable" plastics in municipal leaf compost. *J. Ind. Microbiol.* 10:199–206.

Goodwin, S.D. Unpublished material.

Greizerstein, H.B., J.A. Syracuse, and P.J. Kostyniak. 1993. Degradation of starch modified polyethylene bags in a compost field study. *Polym. Degrad. Stabil.* 39:251–259.

Gu, J.-D., D. Eberiel, S.P. McCarthy, and R.A. Gross. 1993. Degradation and mineralization of cellulose acetate in simulated thermophilic compost environments. *J. Environ. Polym. Degrad.* 1:281–291.

Gu, J.D., S. Yang, R. Welton, D. Eberiel, S.P. McCarthy, and R.A. Gross. 1994. Effect of environmental parameters on the degradability of polymer films in laboratory-scale composting reactors. *J. Environ. Polymer Degrad.* 2:129–135.

Hedges, J.I., R.A. Blanchette, K. Weliky, and A.H. Devol. 1988. Effects of fungal degradation on the CuO oxidation products of lignin: a controlled laboratory study. *Geochim. Cosmochim. Acta* 52:2717–2726.

Hocking, P.J., R.H. Marchessault, M.R. Timmins, T.M. Scherer, and R.W. Lenz. 1994. Enzymatic degradability of isotactic vs. syndiotactic poly (beta-hydroxybutyrate). *Macromol. Rapid Commun.* 15:447–452.

Huang, S.J. 1990. Polymer waste management — biodegradation, incineration and recycling. *Polym. Mater. Sci. Eng.* 63:633–636.

Inbar, Y., Y. Chen, and V. Hadar. 1989. Solid state carbon-13 nuclear magnetic resonance and infrared spectroscopy of composted organic matter. *Soil Sci. Am. J.* 53:1695–1701.

Janshekar, H., C. Brown, and A. Fiechter. 1981. Determination of biodegraded lignin by ultraviolet spectrophotometry. *Anal. Chim. Acta* 130:81–91.

Jarrett, P., S.J. Huang, J.P. Bell, J.A. Cameron, and C. Benedict. 1982. Study of the biodegradation of crosslinked films of polycaprolactone. *ACS Org. Coat. Appl. Polym. Sci. Proc.* 47:45–48.

Kaplan, D.L., R. Hartenstein, and J. Sutter. 1979. Biodegradation of polystyrene, poly(methyl methacrylate) and phenol formaldehyde. *Appl. Environ. Microbiol.* 38:551–553.

Karlsson, S., O. Ljungquist, and A.-C. Albertsson. 1988. Biodegradation of polyethylene and the influence of surfactants. *Polym. Degrad. Stabil.* 21:237–250.

Kavelman, R., and B. Kendrick. 1978. Degradation of a plastic polyepsilon-caprolactone by hyphomycetes. *Mycologia* 70:87–103.

Kawai, F., T. Kimura, M. Fukaya, Y. Tani, K. Ogata, T. Veno, and H. Fukami. 1978. Bacterial oxidation of polyethylene glycol. *Appl. Environ. Microbiol.* 35:679–684.

Kirk, T.K., W.J. Connors, and J.G. Zeikus. 1976. Requirements for a growth substrate during lignin decomposition by two wood-rotting fungi. *Appl. Environ. Microbiol.* 22(1):192–194.

Kirk, T.K. and R.L. Farrell. 1987. Enzymatic "combustion": the microbial degradation of lignin. *Ann. Rev. Microbiol.* 41:465–605.

Komarek, R.J., R.M. Gardner, C.M. Buchanan, and S. Gedon. 1993. Biodegradation of radiolabeled cellulose acetate and cellulose proprionate. *J. Appl. Polym. Sci.* 50:1739–1746.

Kontchou, C.Y. and R. Blondeau. 1992. Biodegradation of soil humic acids by *Streptomyces viridosporus*. *Can. J. Microbiol.* 38:203–208.

Krupp, L.R., and W.J. Jewell. 1992. Biodegradability of modified plastic films in controlled biological environments. *Environ. Sci. Technol.* 26:193–198.

Laishram, I.D., and P.S. Yadava. 1988. Lignin and nitrogen in the decomposition of leaf litter in a sub-tropical forest ecosystem at Shiroy hills in north-eastern India. *Plant Soil* 106:59–64.

Lamar, R.T. and D.M. Dietrich. 1990. In situ depletion of pentachlorophenol from contaminated soil by *Phanerochaete* spp. *Appl. Environ. Microbiol.* 56:3093–3100.

Lee, B., A.L. Pometto, III, A. Fratzke, and T.B. Bailey, Jr. 1991. Biodegradation of plastic polyethylene by *Phanerochaete* and *Streptomyces* species. *Appl. Environ. Microbiol.* 57:678–685.

Lorenz, M.G., and W. Wackernagel. 1987. Adsorption of DNA to sand and variable degradation rates of adsorbed DNA. *Appl. Environ. Microbiol.* 53:2948–2952.

Martinez, A.T., A.E. Gonzalez, A. Prieto, F.J. Gonzalez-Vila, and R. Frund. 1991. *p*-Hydroxyphenyl:guaiacyl:syringyl ratio of lignin in some austral hardwoods estimated by CuO-oxidation and solid state NMR. *Holzforschung* 45:279–284.

Milstein, O., R. Gersonde, A. Huttermann, M.-J. Chen, and J.J. Meister. 1992. Fungal biodegradation of lignopolystyrene graft copolymers. *Appl. Environ. Microbiol.* 58:3225–3232.

Milstein, O., R. Gersonde, A. Huttermann, R. Frund, H.J. Feine, H.D. Ludermann, M.-J. Chen. and J.J. Meister. 1994. Infrared and nuclear magnetic resonance evidence of degradation in thermoplastics based on forest products. *J. Environ. Polym. Degrad.* 2:137–152.

Mullins, C.E., and B.J. Hutchison. 1982. The variability introduced by various sub-sampling techniques. *J. Soil Sci.* 33:547–561.

Narayan, R. 1993. Biodegradation of polymeric materials (anthropogenic macromolecules) during composting, in H.A.J. Hoitink and H.M. Keener (Eds.), *Science and Engineering of Composting,* Renaissance Publications, Worthington, OH, pp. 339–362.

Nishida, H., and Y. Tokiwa. 1993. Distribution of poly (beta-hydroxybutyrate) and poly (epsilon-caprolactone) aerobic degrading microorganisms in different environments. *J. Environ. Polym. Degrad.* 1:227–233.

OECD. 1981. *Guidelines for Testing Chemicals. Section III. Degradation and Accumulation.* Organization for Economic Cooperation and Development, Paris.

Ogram, A.V., M.L. Mathot, J.B. Harsh, J.Boyle, C.A. Pettigrew. 1994. The effects of DNA polymer length on its adsorption to soils. *Appl. Environ. Microbiol.* 60, 393–396.

Osmon, J.L., and R.E. Klausmeier. 1978. Techniques for assessing biodeterioration of plastics and plasticizers, In A.H. Walters (Ed.), *Biodeterioration Investigation Techniques.* Applied Science Publishers, Englewood, NJ, pp. 77–94.

Owen, W.F., D.C. Stuckey, J.B. Healy, L.Y. Young, and P.L. McCarty. 1979. Bioassay for monitoring biochemical methane potential and anaerobic toxicity. *Water Res.* 13:485–492.

Palmisano, A.C., and C.A. Pettigrew. 1992. Biodegradability of plastics. *BioScience* 42:680–685.

Pettigrew, C.A., G.A. Rece, M.C. Smith, and L.W. King. 1995. Aerobic biodegradation of synthetic and natural polymeric materials: a component of integrated solid-waste management. *J. Macrobiol. Sci. Pure Appl. Chem.* A32(4):811–821.

Pitt, C.G., and A. Schindler. 1983. Biodegradation of polymers, in S.D. Bruck (Ed.), *Controlled Drug Delivery,* Vol. 1. CRC Press, Boca Raton, FL, pp. 53–80.

Pohland, F.G., W.H. Cross, and L.W. King. 1993. Codisposal of disposable diapers with shredded municipal refuse in simulated landfills. *Water. Sci. Technol.* 27:209–223.

Potts, J.E., R.A. Clendinning, W.B. Ackart, and W.D. Niegisch. 1973. The biodegradability of synthetic polymers, in J. Guillet (Ed.), *Polymers and Ecological Problems: Polymer Science and Technology,* Vol. 3. Plenum Press, New York, pp. 61–79.

Rausa, R., E. Mazzolari, and V. Calemma. 1991. Determination of molecular size distribution of humic acids by high-performance size-exclusion chromatography. *J. Chromatogr.* 541:419–429.

Reinhammer, B. 1984. Laccase, in R. Lontie (Ed.), *Copper Proteins and Copper Enzymes,* CRC Press, Boca Raton, FL, pp. 1–35.

Robert, D., and C.-L Chen. 1989. Biodegradation of lignin in spruce wood by *Phanerochaete chrysosporium*: quantitative analysis of biodegraded spruce lignin by ^{13}C-NMR spectroscopy. *Holzforschung* 43:323–332.

Rusoff, I.I., R.R. Baldwin, F.J. Domingues, C. Monder, W.J. Ohan, and R. Thiessen. 1960. Intermediary metabolism of adipic acid. *Toxicol. Appl. Pharmacol.* 2:316–330.

Saiz-Jimenez, C., and J.W. De Leeuw. 1984. Pyrolysis-gas chromatography-mass spectroscopy of isolated, synthetic and degraded lignins. *Org. Geochem.* 6:417–422.

Saiz-Jimenez, C., N. Senesi, and J.W. De Leeuw. 1989. Evidence of lignin residues in humic acids isolated from vermicomposts. *J. Analyt. Appl. Pyrol.* 15:121–128.

Scheu, S., S. Wirth, and U. Eberhardt. 1993. Decomposition of ^{14}C-labeled cellulose substrates in litter and soil from a Beechwood on limestone. *Microb. Ecol.* 25:287–304.

Schnitzer, M. 1971. Characterization of humic constituents by spectroscopy, in A.D McLaren and J. Skujens (Eds.), *Soil Biochemistry,* Vol. 2. Marcel Dekker, NY, pp. 60–95.

Schwab, B.S., C.J. Ritchie, D.J. Kain, G.C. Dobrin, L.W. King, and A.C. Palmisano. 1994. Characterization of compost from a pilot plant scale composter utilizing simulated solid waste. *Waste Manage. Res.* 12:289–303.

Seal, K.J. 1988. The biodeterioration and biodegradation of naturally occurring and synthetic plastic polymers. *Biodeter. Abstr.* 2:295–317.

Sharabi, N.E., and R. Bartha. 1993. Testing some assumptions about biodegradability in soil as measured by carbon dioxide evolution. *Appl. Environ. Microbiol.* 59:1201–1205.

Shelton, D.R., and J.M. Tiedje. 1984. General method for determining anaerobic biodegradation potential. *Appl. Environ. Microbiol.* 47:850–857.

Shimp, R.J. 1993. Assessing the environmental safety of synthetic materials in municipal solid waste derived compost, in H.A.J. Hoitink and H.M. Keener (Eds.), *Science and Engineering of Composting,* Renaissance Publications, Worthington, OH, pp. 383–400.

Shimp, R.J., R.J. Larson, and R.S. Boethling. 1990. Use of biodegradation data in chemical assessment. *Environ. Toxicol. Chem.* 9:1369–1377.

Sinsabaugh, R.L., R.K. Antibus, and A.E. Linkins. 1991. An enzymatic approach to the analysis of microbial activity during plant litter decomposition. *Agric. Ecosys. Environ.* 34:43–54.

Smith, S.C, B. Low, and R. Herman. 1992. Design and performance of a rotating drum composter for evaluation of degradability of nonwovens and other materials. *INDA-JNR*, 4:21–25.

Stegmann, R., S. Lotter, L. King, and W.D. Hopping. 1993. Fate of an absorbent gelling material for hygiene paper products in landfill and composting. *Waste Manage. Res.* 11:155–170

Stevenson, F.J. 1994. *Humus Chemistry. Genesis, Composition, Reactions*, Wiley-Interscience, New York.

Sturm, R.N. 1973. Biodegradability of nonionic surfactants: screening test for predicting rate and ultimate biodegradation. *J. Am. Oil Chem. Soc.* 50:159–167.

Suflita, J.M., C.P. Gerba, R.K. Ham, A.C. Palmisano, W.L. Rathje, and J.A. Robinson. 1992. The world's largest landfill: a multidisciplinary investigation. *Environ. Sci. Technol.* 26:1486–1495.

Sykes, M.L., H.W. Yeh, I.F. West, R.W. Gauldie, and C.E. Helsley. 1994. A carbon-13 method for evaluating degradation of starch-based polymers. *J. Environ. Polym. Degrad.* 2:201–209.

Tien, M. and T.K. Kirk. 1983. Lignin-degrading enzyme from the Hymenomycete *Phanerochaete chrysosporium* Burds. *Science* 221:661–663.

Tilstra, L., and D. Johnsonbaugh. 1993. A test method to determine rapidly if polymers are biodegradable. *J. Environ. Polym. Degrad.* 1:247–255.

Tsuchii, A., T. Suzuki, and K. Takeda. 1985. Microbial degradation of natural rubber vulcanizates. *Appl. Environ. Microbiol.* 50:965–970.

Tsuchii, A., K. Hayashi, T. Hironiwa, H. Matsunaka, and K. Takeda. 1990. The effect of compounding ingredients on microbial degradation of vulcanized natural rubber. *J. Appl. Polym. Sci.* 41:1181–1187.

USEPA. 1994. Characterization of Municipal Solid Waste in the United States: 1994 Update. EPA530-R-94-042, U.S. Environmental Protection Agency, Washington, D.C.

USFDA. 1987. Environmental Assessment Technical Assistance Handbook. PB87–175345. National Technical Information Service, U.S. Food and Drug Administration, Washington, D.C.

van der Zee, M., L. Stjtsma, G.B. Tan, H. Tournois, and D. de Wit. 1994. Assessment of biodegradation of water insoluble polymeric materials in aerobic and anaerobic aquatic environments. *Chemosphere* 28:1757–1771.

Wang, Y.-S., C.S. Byrd, and M.A. Barlaz. 1994. Anaerobic biodegradability of cellulose and hemicellulose in excavated refuse samples using a biochemical methane potential assay. *J. Ind. Microbiol.* 13:147–153.

Watanabe, T. 1985. A scanning electron microscopic study of microbial deterioration of polyacrylonitrile(acrylic) textiles. *J. Soc. Fiber Sci. Technol. (Japan)* 41:37–42.

Westermark, U., and K.-E. Eriksson. 1974. Cellobiose: quinone oxidoreductase, a wood-degrading enzyme from white-rot fungi. *Act. Chem. Scand.* B28:209–214.

Williams, G.R. 1982. The breakdown of rubber polymers by microorganisms. *Intl. Biodeter. Bull.* 18:31–36.

Wojtás-Wasilewska, M., J. Luterek, A. Leonowicz, and A. Dawidowicz. 1988. Dearomatization of lignin derivatives by fungal protocatechuate 3,4-dioxygenase immobilized on porosity glass. *Biotech. Bioeng.* 32:507–511.

Yabannavar, A., and R. Bartha. 1993. Biodegradability of some food packaging materials in soil. *Soil Biol. Biochem.* 25:1469–1475.

Yabannavar, A.V., and R. Bartha. 1994. Methods for assessment of biodegradability of plastic films in soil. *Appl. Environ. Microbiol.* 60:3608–3614.

Yakabe, Y., and H. Tadokoro. 1993. Assessment of biodegradability of polycaprolactone by MITI test method. *Chemosphere* 27:2169–2176.

Index

A

Absorbent gel material (AGM), 194

Accelerated methane production phase, 11, 42–44

Acetate formation, in anaerobic decomposition, 37, 39

Acetogenesis in anaerobic decomposition pathway, 37–38

Acetogenic bacteria, 10, 42–43

Acidogenesis,
Monod kinetic model description for, 89–90

Acridine orange direct count (AODC), 48–49, 51

Actinomycetes
in composting materials, 124–125
population trends, 116–118

Adenoviruses, 160

Adipic acid, metabolization of, 194,195

Aerated static pile, 143

Aerobic decomposition, see also Composting
in laboratory-scale reactors, 41
in landfills, 9, 11

Aerobic respirometric analysis of polymer degradation, 187–192

Air pollution, from fossil fuels combustion, 73–74

American Society for Testing Materials (ASTM)
gross molecular analysis, 185–187
incubation temperature, 190
use of bacterial isolates, 181
use of fungal species, 181

Amylase activity measurement, 53

Anaerobic acid phase of decomposition, 41–42

Anaerobic bacteria
methanogenic, 76–79
refuse decomposition in landfills by, 9, 37–39
techniques for study of activity, 48–55

Anaerobic composting, 16

Anaerobic decomposition, general pathways, 37–39

Anaerobic digesters
design, 16, 73, 100–101
hydrolytic enzymes in, 15, 76–80
loading rates in, 94–96
operational parameters, 93–98
performance parameters of, 84, 98–101
predominant hydrolytic bacteria in, 76–79

Anaerobic digestion, 9–16, 37–48
applications and benefits, 72–74, 103
commercialization of, 101–103
final products of, 14
generalized scheme for, 14, 37–39, 72–73
in landfills, 9–16, 37–48
phases in bioreactors, 40–46
polymer hydrolysis in, 9–10, 103
reactor configurations, 16
temperature effects on rate, 16
in digesters, 96–97

Anaerobic digestion process models, 87–93
fluid dynamics, 92–93
kinetics models, 88–91
effects of inhibitors, 90–91
mass balance equations, 88
mass transfer resistance, 91–92

Anaerobic fungi, 77

Anaerobic respirometric analysis, 194–195

Archaea, 81, 83–84

Ash, from MSW combustion, 8

Aspergillus fumigatus, 160

Autoecological investigation of composting ecosystems, 120

B

Bacillus stearothermophilus, 135

Bacteria, see Anaerobic bacteria and Human microbial pathogens

215

Bactivory, 24
Bins for composting, 143
Biochemical Methane Potential (BMP)
    assay, 61, 84, 188, 192–193
    application in digesters, 98–99
    application in landfill, 61
    test of synthetic polymeric material
        biodegradability, 192
Biodegradability of polymeric materials,
    see Synthetic polymer
    biodegradation
Biogas, 73, 102
Biological decomposition of MSW,
    9–18; see also Aerobic
    decomposition and Anaerobic
    decomposition
    in anaerobic digesters, 14–16
    in bioreactors, 39–44
    in composting facilities, 17–18; see
        also Composting of MSW
    in landfill, 9–14
        comparison with laboratory-scale
            system, 45–48
Biological oxygen demand (BOD), of
    leachates, 14–15
Biological self-heating, 121
Biomass measurements, 83–84
Biomethanogenesis in anaerobic
    digestion
    description and applications,72–74
    inhibition, 79, 80, 90–91
    microbiology of, 74–81
        activity measurements, 84–87
        biomass measurements, 83–84
        culturing techniques, 82
        immunobiological and nucleic acid
            probes, 82–83
    operational parameters, 93–98
    performance parameters, 98–101
    process modeling, 87–93
        fluid dynamics, 92–93
        kinetics, 88–91
        mass balances, 88
        mass transfer effects, 91–92
    rate of methane production, 98–99
        measure of digester performance,
            84
    reduction of organic matter, 73
    study methods of, 82–84

Biomethanogenesis pathways of
    anaerobic decomposition, 37–38
Bioreactors
    comparison with landfill
        decomposition, 45, 46–48
    enclosed composting system, 144
    as landfill, 11
    phases of decomposition in, 39–44

C

Campylobacter, 156, 157
Carbon dioxide
    composting product, 17
    concentrations during decomposition,
        41, 42, 44
    greenhouse gas from landfill, 20, 32
Carbon monoxide, as microbial activity
    parameter in anaerobic digesters,
        85
Carboxyl(ic) acid
    chelating effects of, 47
    concentration changes during
        decomposition, 11, 37, 41–44
        dissolution of inorganic constituents
            in leachate, 45, 47
        effects on methane production,
            44
Cellulase, 76
    assay limitation, 54
    enhancement in anaerobic digester,
        78
    in MSW digesters, 95
    measurement of, 53–54
    of *Clostridium thermocellulase,* 76
    synthesis of, 77
^{14}C-Cellulose mineralization, 54–55
C:N ratios, 128–129
    and ammonia release, 137
Cellulase, 76
    assay limitation, 54
    discrete cellulases in MSW digesters,
        95
    synthesis, 77
Cellulase activity
    enhancement in anaerobic digesters,
        78
    measures of, 53–54
Cellulomonas fermentans, 56

Cellulose
  anaerobic digestion of, 76, 96
  major biodegradable landfill substrate,
    12, 15, 36–37
Cellulose-degrading microorganisms,
  76
  in landfill, 12, 37–38
    roll tube enumeration of, 52
  presence in MSW digester, 15
Cellulose hydrolysis in biodegradation of
  MSW, 15
Cellulolytic enzyme system, 76
Chemical contaminants, in MSW,
  5–6
Chemical oxygen demand (COD),
  45
Chlorinated aliphatic hydrocarbons
  (CAHs), 5
Clear zone assays, 187
Closed Bottle test, 187–188
Clostridia, 77
  pectinolytic activity of, 78
CoM, coenzyme in methanogenic
  bacteria, 81
Coliform bacteria, 158
Collection alternatives for MSW, see
  Integrated solid waste
  management
Combustion of MSW
  atmospheric pollution from, 73–74
  estimated percentage of total disposed,
    6, 8
  process in integrated solid waste
  management, 8
Compost
  from anaerobic digestion, 102
  matrix compositional and structural
    effects, 130–131
  maturity, 17–18
    parameters, 119
    humification, 139
  reduction of heavy metal
    contamination, 119
Compost-CO_2 test, 188, 190
Composting microbiology
  actinomycetes, 124–125
  activity and environmental factors,
    122, 132–140
  changes in substrate, 139–140

gas exchange, 135–136
heat evolution during composting,
  134–135
pH and ammonia, 137–139
temperature control by turning
  compost, 134
water, 136–137
autoecological approaches, 120,
  121
biological self-heating, 121
carbon and nitrogen assimilation by,
  128–129
fate of pathogens, 120
fungi, 125–127
nonactinomycete bacteria, 122–124
odorous compounds formation,
  140–142
population trends during process,
  116–118, 122–127, 128–129
  effects of temperature and nutrition
    on, 128–129
  successional nature of, 121–122,
    125
synecological approaches, 120, 122
Composting of MSW, 17–18,
  116–145
application to waste treatment, 8,
  119
description of process, 17, 116
components of, 14–15
curing phase, 117
facilities, 8, 119
factors affecting, 18
fate of pathogens during, 18
microbiology, see Composting
  microbiology
microbial ecology of
  general stages, 116
  interactions with physical and
    chemical environment,
    132–139
nuisance odors, 140–142
  common compounds producing,
    18
phases of, 17–18, 116–118
process phases, 17
processing goals and problems, 119,
  120
process management, 142–145

products of, 17
separation of materials, 130
    contamination issue, 119
sources of sulfur in, 18
stages, 116–118
substrate changes, 139–140
temperature effects, 116
    ventilation, 116, 120
volatile emissions from, 18
Confocal laser microscopy, 24
Continuously stirred tank reactor
    (CSTR), 16
Coxsackieviruses, 159, 163
Cross polarization and magnetic-angle
    spinning (CPMAS), 200
Cryptosporidium, 158–159, 164
Curie-point pyrolysis-gas
    chromatography mass
    spectrometry, 201
Curing, phase in composting, 117

D

Decelerated methane production phase,
    11, 44
Decomposition of MSW, see also
    Aerobic decomposition and
    Anerobic decomposition
    in bioreactors, 39–44
    comparison with landfills, 45–48
    in composting, see Composting
    economic benefits of, 3
    laboratory-scale landfill models or
        lysimeters, 9
    in landfill, see Landfill
Denaturing gradient gel electrophoresis
    (DGGE), 24
Depolymerization in anaerobic
    digestion, 74–79
Digesters, see Anaerobic digesters
Dihydrogen fermentation product in
    anaerobic digestion, 79, 80
Dinitrification, and aerobic
    decomposition, 11
Disposal alternatives for MSW, see
    Integrated solid waste
    management
Disposable diapers, source of pathogens
    in MSW, 163–164

Domestic pet waste, source of pathogens
    in MSW, 162–163
Drinking water contamination, see
    Groundwater contamination
Dry anaerobic digestion, 16

E

Energy recovery from MSW
    combustion, 8
    economic value of, 3
    fuel from anaerobic digestion, 103
    landfill methane, 20, 32
Enteric bacteria, see Human microbial
    pathogens in MSW
Enteric pathogens, see Human microbial
    pathogens in MSW
Enteric viruses, see Human enteric
    viruses
Enteroviruses, 159
Enzyme assays of polymer
    biodegradability, 182–184
Environmental consequences of MSW,
    18
    benefits of anaerobic digestion,
        103
    air pollution, 73–74
    water contamination from leachates,
        14, 18–19
Escherichia coli, 156, 157–158
Esterase activity, measurement of, 53

F

F_{420} coenzyme
    in methanogenic bacteria, 81
    estimation of methanogenic biomass
        by, 12
    identification and enumeration of
        methanogens by fluorescence of,
        81, 82, 85–86
Fatty acids in MSW, anaerobic digestion
    of, 79
Fecal indicator bacteria, 156
    in leachate, 19
Feedstock characteristics of MSW for
    anaerobic digestion, 93–94
Fermentation
    activity in anaerobic digestion, 79–80

in general pathways of anaerobic
decomposition, 37–38
of high solids, 16
Fermentative bacteria
enumeration techniques for, 51–52
role in anaerobic decomposition,
37–38
role in MSW degradation in landfills,
10
Field- or full-scale confirmation tests of
polymer biodegradation, 180,
201–203
in solid waste treatment and disposal,
201
in terrestrial ecosystems, 202
First order kinetic model of anaerobic
digestion, 89, 90
Fourier transform infrared spectroscopy,
198–199
DRIFT, 199–200
Fungi
in composting materials, 125–127
landfill anaerobic pathogens, 160
population succession in MSW,
116–118, 120–121, 122, 124
role in biological self-heating, 121

G

Gas component analyses
BMP assay, 98–99
evaluating microbial activity and
digester performance, 84–85
Gastroenteritis, 156
from adenoviruses, 160
from *Cryptosporidium* and *Giardia*,
159
Norwalk virus and, 160
from rotaviruses, 160
strains of *E. coli* causing, 158
Gel permeation chromatography (GPC),
197–198
Giardia, 158–159
Giardiasis, 159, 164
Greenhouse gases, see also Carbon
dioxide and Methane
from landfills, 20, 32, 36
Gross macromolecular analysis,
185–187

Groundwater contamination, 9, 19–20
landfill monitoring and prevention of,
36, 63

H

Hazardous waste in MSW, 5
Heavy metals in MSW, 5–6
Hemicellulase, in anaerobic digestion,
74–78
Hemicellulose
anaerobic degradation, 77–78
structural barriers to, 76
in MSW, 36–37, 76
Hepatitis A virus, 159–160
Hepatitis B virus, 159–160
Household hazardous products, 5
Human enteric viruses, 159–160
in laboratory MSW lysimeters,
166
in landfilled disposable diapers,
19–20, 164–165
Human microbial pathogens in MSW
attentuation during composting,
168–170
enteric bacteria, 156–158
enteric protozoa, 158–159
enteric viruses, 159–160
fecal indicator bacteria and water
contamination by, 156, 158
fungi, 160
major pathogens, 156–157
occurrence and survival in landfill,
19–20, 156, 164–166
occurrence and survival in leachate,
19–20, 166–168
sources of, 19, 161–164
Humification, 139
Hydraulic retention time, 94, 96
Hydrogen gas, measurement of microbial
activity and digester
performance, 84–85
Hydrolases, 74–75
Hydrolysis, see also
Depolymerization
differentiated from depolymerization,
74
in general pathways of anerobic
decomposition, 37–38

Hydrolytic enzymes, see Microbial
    hydrolytic enzymes
  in anaerobic digestors, 15
Hydrolytic microorganisms
  in anaerobic digesters, 16

                    I

Immunological probes, 82–83
Integrated solid waste management,
    6–8
  life-cycle inventory parameters,
    22–23
  role of landfills, 8, 32
Interspecies acetate transfer, 79–80
Interspecies hydrogen transfer, 79–80
Isotopically labeled materials and
    biodegradability, 193–195

                    L

Laboratory and pilot-scale tests for
    biodegradability, 180, 193–194
Laboratory-scale landfill (lysimeter),
    9–11
  fate of indicator bacteria in, 164
  leachate recycling and neutralization
    in, 11
Landfill
  decomposition in, 9–14, 37–55
    anaerobic bacteria involvement,
      48–55
    comparison of laboratory to
      full-scale results, 45–48
    components carrying refuse-
      decomposing microbes into
      landfill, 50
    end products of, 38
    factors limiting, 61–63
    general pathways, 37–39
    organic components of, 36
    phases in bioreactors, 39–44
    principal biodegradable substrates
      of, 36
    visual categories of components, 36
  design and operation of, 32, 33–36,
    63
    regulations concerning, 34
    typical capacity of, 34

estimated numbers in U.S., 33
environmental impact of, 9, 19–20
gases, see Landfill gases
leachate, see Landfill leachates
methane production from, 10, 11,
    32–33
  alternative management of, 20
  ramifications of, 32
  regulatory policies influencing,
    63
microbiology of, 9–14, 37–39
  cellulolytic and other hydrolytic
    bacteria, 37–38
  enteric pathogens, 161–164
  measurements of microbial
    populations and activities, 48–55
  research, 55–59
  systems for study, 60–63
model scale, 9–11
percent of total MSW generated
    disposed in, 6, 8
role in MSW management, 8
Landfill gas, 32–33; see also Carbon
    dioxide and Methane
  alternatives for management of,
    20, 36
  toxic trace organics in, 5
Landfill leachates, 14–15, 32–36
  composition, 20
    chemical contaminants in, 5–6
    variation in, 52
  effect of methane production on
    strength of, 32
  enteric pathogens in, 164–168
  parameter changes, 15
  public health and environmental
    concerns, 19–20, 63
  quality monitoring, 36
  recycling and neutralization, 62–63,
    64
Landfill liner system, 34
Leachate recycling and neutralization,
    11, 45, 62–63, 64
Leachates, 14
  enteric viruses in, 19
  fluid dynamics in anaerobic digestion,
    92–93
Life-cycle inventory (LCI) parameters,
    23

Lignin
in MSW, 36–37, 75–76
as biodegradation benchmark,
207
depolymerization, 78, 183–184
Lignin peroxidase, 183
Lignocellulose, 74–74
as biodegradation benchmark, 205
Lipase activity, measurement of, 53
Lipid solubilizing anaerobes, 79
Liquid inoculum formation from refuse,
49–51
Loading rate parameters for MSW
digesters, 94–96
Lyases, 75

M

Mass transfer resistance in anaerobic
digestion process model, 91–92
Materials recovery facilities (MRFs),
6–8
Matric phase ecosystem in compost,
130–132
Mesocosm systems of analysis, 195–196
Mesophilic microbial population trends
during composting, 117–118
Methane, atmospheric
emissions sources, 20, 21, 72
Methane gas content
measurement of digester performance,
84
yield in biogas from typical
OFMSW plant, 102
Methane production in landfill, 10–11
benefits of, 32–33; see also Energy
recovery
biological decomposition time range,
39
contribution to greenhouse gases, 20,
32
major factors limiting
moisture and pH, 11
cellulose and hemicellulose content
in MSW, 36, 37
microbiological mediation of, 9–10
regulatory factors influencing, 20
Methane production phases in
bioreactors, 39–44

Methane yield
depolymerization efficiency and, 75
from decomposition in reactors,
63–64, 94
effects of sulfate on, 64–65
Methanogen coenzymes, 81
Methanogenesis, see also
Biomethanogenesis
Monod kinetics model description of,
89–90
Methanogenic bacteria, 12–14, 41,
80–81
in anaerobic digestion, 39
role in landfill MSW decomposition,
10
Methanogenic reactions, 81
Metopus palaeformis, 56–57
Microbial activity measurements
in anaerobic digesters, 84–87
in landfill, 48–55
methane production, 54
neutralization of ^{14}C-cellulose,
54–55
Microbial hydrolytic enzymes
in anaerobic digesters, 15
activity measurement, 86–87
and design of biodegradable polymers,
25
Microbial colonization assays,
182–184
Microbial substrate utilization in
anaerobic digestion, 89–91
Microcalorimetry, applied to anaerobic
digesters, 87
Microorganisms associated with
decomposing MSW
in composting process, 3
in anaerobic digesters, 3
diversity of, 3
in landfill, see Landfill
methods of community analysis,
23–24
pathogenic, see Human pathogenic
microbes in MSW
population changes during
decomposition in bioreactor,
42
refuse surface analysis technology,
24–25

surface polymeric biodegradation,
     182–184
Mineralization of straight chain
     hydrocarbons, 197
Mixed culture screening level tests,
     185–193
   aerobic respirometric analysis,
     187–192
   anaerobic respirometric analysis,
     192–193
   clear zone assays, 185
   gross macromolecular analysis,
     185–187
Monod kinetic model, 89–90
Most probable number (MPN) tests, 48
Municipal solid waste (MSW)
   biodegradable fraction, 4, 14
   composition, 3–6, 9, 36–37, 93–94
     organic, 36–37
   definition of MSW, 4
   economic benefits from
     decomposition of, 3
   environmental threats from, 18
   feed characteristics for anaerobic
     digestion, 93–94
   generation per capita in U.S., 6
   heterogeneity of, 2–3, 93
   human health concerns related to,
     19–20; see also Human microbial
     pathogens
   management processes, 6
   microbial colonization of, 2–3
Municipal solid waste decomposition,
     see Biological decomposition of
     MSW

N

Nitrification inhibitors, 188
Nonactinomycete bacteria, 116, 122–124
Norwalk virus, 160
Nucleic probes, 82–83
Nuclear magnetic resonance (NMR), 200

O

Obligate proton-reducing acetogenic
     bacteria, 38–39
OECD 301C method, 188–189

Oligonucleotide probes, 83, 84
Organic acids, mineralization of, 10
Oxidation-reduction potential, as
     anaerobic digester performance
     parameter, 100
Oxygen, depletion in MSW
     decomposition, 41

P

Paper and paperboard in MSW, 4
   methane yield from, 63–64
Pathogenic microorganisms, see Human
     microbial pathogens in MSW
Pectins, depolymerization of, 78
Petroleum hydrocarbons, 5
Phanerochaete chrysosporium, 127
pH inhibition factor in
     biomethanogenesis, 90–91
   parameter of anaerobic digester
     performance, 99–100
Plant tissue components of MSW,
     75–76
Plastics, 4, 21–22; see also Polymer
     biodegradability
Polyolefin derived plastics, 21–22
Poliovirus, 20
Polyethylene, biodegradable alternatives
     for, 22
Polypropylene, biodegradable
     alternatives for, 22
Polymer, properties of, 177, 179
Polymer biodegradability, see Synthetic
     polymer biodegradation
Polymer hydrolysis, see also Anaerobic
     decomposition
   in landfill, 9–11
Product life-cycle analysis, 22–23
Proprionate, formation in anaerobic
     digestion, 79
Protease activity, measurement of,
     53
Protein
   anaerobic digestion of, 78–79
   concentrations in MSW, 36
Protozoa, in landfill, 14, 19
Pure culture screening level test,
     180–185
   enzyme assays, 182–184

microbial colonization assays,
180–182
respirometric analysis with pure
cultures, 184–185

R

Reactor configurations, 16
MSW anaerobic digesters,
100–101
Recycling of MSW components, 6–8
Reduced NAD(P)
measurement of methanogenic activity
in anaerobic digesters, 85–86
Reductive dehalogenation, 5
Refuse, see Municipal solid wastes
Refuse-derived fuel (RDF), 8
Residue and degradation product
analysis, 198–203
Respirometric analysis using pure
cultures, 184–185
16S rRNA probes, 24, 83
Rotaviruses, 160

S

Salmonella, 156, 157
Sanitary landfills, see Landfills
Scanning electron microscope,
24–25
assessment of synthetic polymeric
material surfaces, 186–187
Separation alternatives for MSW, see
Integrated solid waste
management
Sewage sludge, source of pathogens,
161–162
Shigella, 156–157
Shrinking core kinetic model, 89,
90
Silo configuration for composting,
144
Soil amendment, 3
contamination prevention, 6
Soil-CO_2 test, 191
Solid state fermentation, 16
Solid waste, 6; see also Integrated solid
waste management
Soluble sugars, in MSW, 36

Solubilization during anaerobic
digestion, 75
kinetic model for, 89, 90
Starch in MSW, anaerobic digestion of,
78
Step diffusional kinetic model, 89, 90
Sturm test, 189–190
Sulfate in MSW, 64–65
effects on decomposition and methane
production, 65
Sulfate-reducing bacteria, role in landfill,
10
Synecological investigation of
composting ecosystems, 120
Synthetic polymer biodegradation,
21–22, 25
benchmarks, 205
environmental factors affecting, 177
glossary of terms, 176–177
macro-and molecular scale studies of,
177–178
standardization of test methods, 178,
203–207
tests for synthetic polymeric materials,
22, 176–203
field- or full-scale tests, 201–202
laboratory- and pilot-scale tests,
193–201
mixed culture screening level tests,
185–193
pure culture screening level tests,
180–185
tiered testing approach for evaluation
of, 179–180
Syntrophs, 79

T

Temperature effects on biodegradation,
16
Thermophilic microbial changes during
composting, 117–118
Thermophilic fungi, changes during
composting, 117–118
Three-stage sequential batch reactor,
16
Tiered testing approach, for synthetic
polymer degradation, 22,
179–180

Tipping fees for anaerobic digestion,
     101–102
Toxic chemicals in MSW, 5–6
Trace organic compounds, 5–6
Treatment alternatives for MSW,
     see Integrated solid waste
     management
Trough systems configuration for
     composting, 143–144
Tunnel configuration for composting,
     144
Two-phase digestion, 16

U

U.S. Environmental Protection Agency
     (USEPA)
   definition of MSW, 4
   landfill greenhouse gases regulation,
     20

V

Vinyl chloride, in landfill gas, 5
Viruses in MSW, see also Human
     microbial pathogens in MSW
   lysis by bacteriophage, 24

Volatile fatty acids, parameter
     influencing digester
     performance, 99–100
Volatile organic chemicals, 16
   in waste collection vehicle headspace,
     5
Volatile solids (VS)
   in MSW components, 37
   reduction in anaerobic digestion,
     102
Volatilization, 5
Volume reduction of MSW
   by composting, 8
   by recycling, 6

W

Weight loss measurements, 185–186
Windrows, 142–143

X

Xenobiotic chemicals, fate in landfill, 19

Y

Yard trimmings, 4, 6